John Berkenhout

Outlines of the natural history of Great Britain and Ireland

Vol. 3

John Berkenhout

Outlines of the natural history of Great Britain and Ireland
Vol. 3

ISBN/EAN: 9783337174828

Printed in Europe, USA, Canada, Australia, Japan

Cover: Foto ©ninafisch / pixelio.de

More available books at **www.hansebooks.com**

OUTLINES

OF THE

NATURAL HISTORY

OF

GREAT BRITAIN

AND

IRELAND.

CONTAINING

A systematic Arrangement and concise Description of all the Animals, Vegetables, and Fossils, which have hitherto been discovered in these Kingdoms.

By JOHN BERKENHOUT, M. D.

IN THREE VOLUMES.

VOL. III.

Comprehending the FOSSIL KINGDOM.

LONDON:

Printed for P. ELMSLY (Successor to Mr. VAILLANT) facing Southampton-Street, in the Strand.

M DCC LXXII.

PREFACE.

IF our young Naturalists owe me any obligation, and if that obligation be in proportion to my trouble, they are more indebted to me for this small volume than for the other Two. In my progress through the Animal and Vegetable Kingdoms I was happy in a faithful, skilful guide; the great Linnæus led the way, and I knew him too well ever to lose sight of him. But, on entering the Fossil Kingdom, I found him not delighted with the country: We differed about the road, and so shook hands. How sorry I was to part with such a companion may be easily imagined, from the pleasure and advantage I must have reaped in the former part of my journey.

Thus deprived of my old friend, I joined company with travellers of different nations,

nations, particularly with Mr. Cronſtedt and Mr. Wallerius, two Swediſh gentlemen of great abilities, our indefatigable Doctor Woodward, Mr. Da Coſta, and ſome others. I gathered ſome information from each; yet our diſputes were ſo frequent, that I was at laſt obliged to bid them farewell, and purſue my own path. Nevertheleſs, I acknowledge myſelf much obliged to Mr. Cronſtedt, and ſhould have been glad of Mr. Da Coſta's company a little farther: He took his leave abruptly.

The Foſſil Kingdom remains yet in a great meaſure uncultivated, owing chiefly to the ſlow progreſs of Chemiſtry, which, till very lately, was but little underſtood; and which, even now, is well underſtood but by very few Naturaliſts. The Reader will probably be ſurpriſed that I have reduced the *Genera* to ſo inconſiderable a number; nevertheleſs, if my Characteriſtics be ſufficiently obvious, he will find the ſtudy of Foſſils by this means greatly facilitated, and my book, conſequently, not intirely uſeleſs.

With

PREFACE.

With regard to the Mineral Waters, it is neceſſary to inform the reader, that, in general, I have copied the authors which are moſt eſteemed. Theſe writers, however, appear to have been very deficient in chemical knowledge; except Dr. Falconer on the Bath Waters, and Dr. Lucas of Dublin. The latter is not infallible. Therefore, concerning the contents of theſe Waters, I do not report them as they really are, but as what our writers on this ſubject have ſuppoſed them to be; which ſuppoſitions, I am certain, are frequently wrong, being contrary to the eſtabliſhed laws of chemiſtry. This ſubject merits further enquiry.

Authors frequently quoted in this Volume.

Lin. i. e. Linnæi Syſtema Naturæ, tom. III. Holmiæ. 1768.

Syſt. Nat. Linnæi Syſtema Naturæ, tom. I. Holmiæ. 1766.

Wall. Wallerius. Mineralogie, traduit de l'Allemand. Paris. 1759.

Cronſt. Cronſtedt's Mineralogy, publiſhed by Da Coſta. 1770.

Falconer. Dr. Falconer on Bath Waters. 1770.

Woodw. Dr. Woodward's Catalogue of Engliſh Foſſils. 1728.

———— ———————— Method of Foſſils. 1728.

Brand. Foſſilia Hantonienſia, a Guſtavo Brander. 4to. 1766.

Smith. Smith's Natural Hiſtory of the Counties of Cork, Kerry, Waterford, 1750. &c.

Luid. Edw. Luidii Lithophylacii Britannici Ichnographia. Oxon. 1760.

Rutty. Synopſis of Mineral Waters. 4to.

Short. Hiſtory of Mineral Waters. 2 vol. 4to. 1734. 1740.

Monro. Treatiſe on Mineral Waters, by Donald Monro, M. D. 2 vol. 8vo. 1770.

Allen.

Allen.   Natural History of the Mineral Waters of Great Britain, by Benj. Allen, M. B. 8vo. 1711.

Shaw.   Enquiry into the Contents, &c. of Scarborough Spaw, by Peter Shaw, M. D. 8vo. 1734.

Hillary.   Enquiry into the Contents, &c. of Lincomb Spaw water, by W. Hillary, M. D. 8vo. 1742.

Pharm. Med.   Pharmacopoeia Medici. edit. altera. 1768.

Vol. I.   Volume the first of this work.

I Divide the Fossil Kingdom into Six Claſſes; viz.

I. EARTHS.

Inſipid, not ſoluble in pure water or oil; not inflammable, not ductile; preſerving their conſtitution in a ſtrong heat, but when fuſed become glaſs.

II. SALTS.

Sapid, ſoluble in water, and recoverable, by evaporation, in their priſtine angular form; not inflammable, nor ductile.

III. INFLAMMABLES.

Soluble in oil, but not in water; readily take fire, and are in a great meaſure conſumed.

IV. METALS.

Of all mineral bodies moſt ponderous; fuſible, but reſuming their original properties and a convex ſurface when cold, even after calcination, by the addition of phlogiſton.

Order I. *Metals* (properly ſo called) malleable.

II. Semi-metals: not malleable.

V. PE-

V. PETREFACTIONS.

Animals, or parts of animals, or vegetables changed into a foſſile ſubſtance.

VI. WATER.

Not inflammable; very little if at all, compreſſible; ſoluble in air; by heat rendered volatile, and extremely elaſtic; fluid when Fahrenheit's thermometer ſtands above 32 degrees, and ſolid when it is below that point.

THE FOSSIL KINGDOM.

FOSSILS are natural bodies unorganized, and without sensation, formed beneath the surface of the earth.

CLASS I.

EARTHS.

Insipid, not soluble in pure water or oil; not inflammable, not ductile; preserving their constitution in a strong heat, but when fused become glass.

> HUMUS. *Mould.* Soft, but not viscid, nor tenacious when moist. Particles heterogeneous, palpable, harsh, diffusible in water. Not soluble in acids. Burns not to lime. Hardens not in the fire.
>
> 1. Communis. *Common Earth.* Covers generally the surface of this Globe. Blackish, or brown. *Wall.* 9. *Lin.* 54, 2. *Cost.* 115. *Wall.* 267.
>
> ARGILLA. *Clay.* Particles impalpable. Viscid and plastic when moist.

Does

Does not effervesce with acids. Hardens in the fire. When pure not vitrifiable.

1. Apyra. *Porcelain Clay.* White, pure, dry, not fusible; burns white, and becomes so hard as to strike fire with steel. Isle of Wight, &c. Cloyne, county of Cork. *Smith*, vol. I. p. 363. *Lin.* 52, 1, 3. *Cronst.* 87. *Cost.* 33.

2. Figulina. *Potters Clay,* or *Pipe Clay.* Mixt with phlogiston and other substances in small proportion. Unctious, white, or bluish, or ash-colour, or black. Burns white, or pearl-colour. In various parts of these kingdoms. *Lin.* 52, 2. *Cost.* 30, &c. *Cronst.* 87. *Woodw.* tom. I. p. 4.

3. Communis. *Common Clay.* Ash-colour, or red, or yellow, or brown, or bluish. Burns red; melts in the fire to a greenish glass. Used for making bricks, tiles, &c. in various parts of these kingdoms. *Lin.* 52, 9. *Cronst.* 97. *Cost.* 30, &c.

4. Bolus. *Bole.* Melts in the mouth; contains a large proportion of iron; grows blackish in the fire, and is then attracted by the magnet.

Red. Norfolk, &c. *Cost.* 11, 86.

Green. Yorkshire, Devonshire, Cornwall. *Cost.* 28. *Cronst.* 94. *Lin.* 52, 13.

Grey.

Grey. In most of our collieries between the beds of coal. *Lin.* 6, 3. *Cronst.* 98.

5. Fullonica. *Fuller's Earth.* Greenish-brown, compact, unctious, glossy when scraped or cut, froths and falls to pieces in water, burns brown and hard. Bedfordshire, Surry, Kent. *Lin.* 5, 2, 7. *Cost.* 69. *Cronst.* 92.

6. Tripolitana. *Tripoli,* or *Rotten Stone.* Brown or yellowish; light, dry, harsh; colours the fingers, breaks easily in the mouth, acquires but little additional hardness in the fire. Used for polishing, &c. Derbyshire; also near Cork in the river Lee, and in Glanmire river near the Bluebell. *Smith,* 382. *Cost.* 87. *Lin.* 52. *Cronst.* 96.

7. Indurata. *Soap-stone.* Hard, compact, unctious, white, often veined with purple or red, or sometimes with green. Not diffusible in water; marks a white line on boards, &c. Near the Lizard-point in Cornwall. *Cost.* 36. *Lin.* 6, 4. *Cronst.* 89.

CALX. *Lime Earth.* Soluble in acids, with the nitrous most perfectly, with effervescence. Being burnt, grows hot with water, and falls to a white powder. Not vitrescent *per se.* Does not strike fire with steel. With borax melts in the fire

fire to a transparent colourless glass. Breaks in any direction.

1. Creta. *Chalk*. Compact, dry, harsh, friable, white, insipid, diffusible in water, adheres to the tongue and fingers. In large strata in the southern counties of England. *Lin.* 53, 1. *Cronst.* 12. *Cost.* 77.

Powder-chalk, or *Mineral Agaric*, or *Lac Lunæ*. A fine light, white powder, differing from common chalk in want of cohesion. Found in the fissures of stone quarries in Oxfordshire, Northamptonshire, &c. *Lin.* 53, 2. *Cronst.* 12. *Cost.* 83. *Woodw.* tom. I. p. 233.

2. Marmor. *Marble*. Hard, solid, opaque, naturally rough, but capable of a fine polish, unless decayed in the air: particles invisible, impalpable. *Lin.* 2, 2. *Cronst.* 13.

Yellow. Monmouthshire, &c. and near Mitchels-town, in the county of Cork. *Smith*, vol. II. p. 376. *Cost.* 197.

Black. With white veins and spots. Broad Helmston and Torbay in Devonshire. *Cost.* 201.

.... With white sea-shells, chiefly turbinated. Near Kilkenny in Ireland, in many parts of which kingdom it is used for chimney-pieces; also near Church-

Church-town, Cork. *Smith*, vol. II. p. 375.

Black. Or bluish, intermixt with spar. Derbyshire. *Short. Min. Wal.* vol. I. p. 24. *Coft.* 149.

.... Without any admixture of white. Caldy Island, Wales. *Woodw.* tom. II. p. 6. Also near Church-town, Cork. *Smith.* vol. II. p. 372.

White. Veined and clouded with deep red. Near Plymouth, Devonshire. *Coft.* 206.

.... Streaked with pale-red. Near Kilarney, in Kerry, Ireland. *Coft.* 206.

.... Glittering with spar, sometimes containing shells. Derbyshire, Wales, &c.

.... Variegated with purple, brown, yellow, or green. In several parts of the county of Cork. *Smith*, vol. II. p. 375. *Coft.* 223.

Grey. Glocestershire, Monmouthshire, &c and at Carigaline, near Cork. *Smith*, vol. II. p. 376. *Coft.* 194.

.... With sparks of spar and flakes of entrochi. Near the lead-mines in Derbyshire. *Coft.* 153.

.... With small white specks. Four miles west of Mallow, in the county of Cork. *Smith*, vol. II. p. 377.

.... With white spots and veins. At Castlemary, in the county of Cork. *Smith*, vol. II. p. 377. *Coſt.* 210.

Grey. With large veins of white ſpar. Near Cork, in the road to Black-rock.

.... Greeniſh, with innumerable turbinated ſhells. *Woodw. Cat.* A. x. b. 60. *Coſt.* 235.

.. Yellowiſh, with innumerable ſmall bivalve ſhells. Whichwood foreſt, Oxfordſhire. *Coſt.* 237.

Blue, frequently veined with white. Near Cork, in the road to Paſſage. Seems to be exactly the ſame marble with that generally uſed in Holland for ſteps on the outſide of their houſes, which they import from Namur, &c. *Coſt.* 198.

Lead colour, clouded with dark ſpots, and white or brown, or both. Kilcrea, eight miles from Cork; alſo at Black-rock. *Smith*, vol. II. p. 378. *Coſt.* 224.

.... With a purple tinge, variegated with yellow, white, aſh-colour, and purple. Whetton, in Derbyſhire. *Coſt.* 224.

Red. Pale-red, glittering, ſparry, interſperſed with black mica. *Coſt.* 157.

Brown. Dull reddiſh brown. Near Aſhburn, in Derbyſhire. *Coſt.* 157.

.... With white and red or purple veins. Devonſhire and Cornwall. *Coſt.* 224.

Brown. With

Brown. With innumerable entrochi, &c. Derbyshire, in great abundance. *Woodw. Cat.* A. x. b. 61. *Coft.* 235.

.... Dark, with numerous semicircular white streaks, the edges of shells. Near Leith, in Scotland. *Coft.* 237.

3. Lapis. *Limeſtone.* Texture less compact, not admitting a fine polish; particles visible, granulated, or scaly. Colour various. In many parts of the three kingdoms. *Lin.* 2, 6. *Cronſt.* 15. *Coft.* 152, &c.

4. Spatum. *Spar.* Breaks eaſily into rhomboidal, cubical, or laminated fragments, with polished surfaces. *Cronſt.* 17, &c.

Common Spar. Rhomboidal, diaphanous or opaque, of various colours. In mines, &c. in Wales, Derbyshire, &c. Ovens, near Cork. *Smith*, vol. II. p. 381. *Woodw.* tom. I. p. 151, &c.

Refracting Spar. Rhomboidal, shews objects seen through it double. In the lead mines in Derbyshire, Yorkshire, &c. *Ibid.*

Chryſtallized Spar. Diaphanous, pyramidal, or columnar. In mines, quarries, caverns, &c. in various parts of these kingdoms. *Woodw.* tom. I. p. 156.

Stalactitical Spar, Iſicle, or *Drop-ſtone.* Formed by the running or dropping of water containing a large proportion of calcareous earth. Opaque, generally laminated, and in various forms from accidental circumstances. Knaresborough in Yorkshire, Ovens near Cork, &c. *Smith*, vol. II. p. 381. *Woodw.* tom. I. p. 155.

MARGA. *Marle.* Calcareous earth mixt with clay. Effervesces with acids whilst crude, but not after being burnt. Hardens in the fire in proportion to the proportion of clay it contains. Vitrescent *per se* into grey glass.

1. Friabilis. *Common Marle.* Diffusible in water. Brown, or brownish. Used for manure in many parts of England. *Cronst.* 31 *Cost.* 67. *Smith*, vol. II. p. 367.

2. Indurata. *Stone-marle.* In loose pieces, grey, or white. In the beds of rivers. *Cronst.* 32. *Cost.* 136. *Lin.* 49, 1.

GYPSUM. *Gyps.* Calcareous earth saturated with vitriolic acid; therefore does not effervesce with acids. Being heated falls to powder, which with water forms plaister, but without exciting heat during the mixture. Melts in the fire *per se*, though sometimes with difficulty, into a white glass. Melted with borax, puffs and bubbles much. Burnt with phlogiston, smells of sulphur; and thus decomposed, or by alkaline salts, its earth is found to contain some iron.

1. Pul-

EARTHS.

1. Pulverulentum. *Plaister Earth.* A dry, harsh, gritty, yellowish white powder; adheres to the tongue, but not to the fingers. Mixt with water forms plaister without previous heating. Clipston quarry, in Northamptonshire. *Woodw.* tom. I. p. 7, 40. *Cronst.* 23. *Lin.* 53, 6. *Cost.* 80.

2. Alabastrum. *Alabaster.* Solid; particles visible, glittering. Less hard than marble, but capable of being polished. Not always saturated with acid, and therefore in some measure soluble in *aqua fortis.* White, or reddish, or yellowish. Derbyshire, Cornwall, Somersetshire, &c. *Cronst.* 23. *Lin.* 3, 3.

3. Commune. *Plaister Stone.* Texture scaly, or granulated; generally red or white; much softer than alabaster, and incapable of a polish. Derbyshire, Yorkshire. *Cronst.* 24. *Lin.* 3, 2.

4. Fibrosum. *Fibrous Plaister Stone.* Called fibrous talc. White. In plaister pits in Derbyshire, Nottinghamshire, &c. *Cronst.* 24. *Lin.* 4. 1. *Cost.*

5. Spatosum. *Selenites.* Diaphanous, rhomboidal, laminated. Frequent in clay-pits, Oxfordshire, Isle of Sheppy, Nantwich in Cheshire, &c. *Cronst.* 24. *Lin.* 16. *Woodw.* tom. I. p. 67. tom. II. p. 10.

TALCUM. Will not strike fire with steel, nor effervesce with acids. Not diffusible in water. In the fire very refractory

fractory *per se*, but fusible with borax. Particles palpable, visible, divisible, flexible, elastic.

1. Mica. *Glimmer.* Particles laminated, scaly, shining, separable, friable, semingly unctious.

Brown. With other fossils in the composition of granites, &c.

Black. In masses of 4 or 5 inches diameter, in rivers in Yorkshire. In gravel-pits in Northamptonshire. *Woodw.* tom. I. p. 63, 64. tom. II. p. 8.

Gold-colour. In small masses or in stones on the sea-shore. *Woodw.* tom. I. p. 63.

Silver Mica. In masses and among the sand on the sea-shore. *Woodw.* tom. I. p. 61.

Greenish. In small masses on Mendip hills, and on the shores in Yorkshire. *Woodw.* tom. I. p. 62.

Grey. On the sea shore. *Woodw.* tom. I. p. 62. 64.

Reddish. On the shores of Lincolnshire, &c.

2. Asbestus. *Asbest.* Surface dull, uneven; texture fibrous. *Cronst.* 112.

Mountain Flax. Grey; filaments long, silky, roundish, straight, intire. In the Highlands of Scotland on the surface of the earth. *Lin.* 7, 1.

Mountain

EARTHS.

Mountain Leather. White, fibres broad, membranous. Highlands of Scotland.

Greenish Asbest. Filaments interrupted or interwoven. In strata of marble in the Isle of Anglesea, &c. *Lin.* 7, 10. *Woodw.* tom. I. p. 77.

SAXUM. Opaque, rough, coarse; particles visible, heterogeneous; not laminated, breaking freely in any direction; will not strike fire with steel, nor burn to lime; not soluble in acids.

1. Cos. *Sand Stone*, or *Free Stone*. Particles small, consisting chiefly of quartz and mica, cemented by clay.

Portland Stone. White, or light grey, of a roundish grit, and glittering. Dorsetshire. Brought frequently to London, and used in building. *Cost.* 125. *Woodw.* tom. I. p. 16.

Purbeck Stone. Brownish ash-colour. Hard, heavy; texture not very compact; composed of an angular grit. Dorsetshire. Used in London for building, &c.

Ketton Stone. Pale-brown, composed of incrusted granules. Heavy, and not very compact. Several Colleges in Cambridge are built with this stone. *Cost.* 129.

Scythe stone. Pale brown, heavy, glittering, and in some degree friable. Used

Used for whet-stones. Derbyshire, &c. *Cost.* 133. *Woodw.* tom. I. p. 17.

Scotish Stone. Bluish, granulated, hard, heavy. Used for new-paving the streets of London. In the king's park near Edinburgh, &c.

Red Sand Stone. Coarse, deep brown-red, friable. Composed of crystalline grit, cemented by a ferruginous earth, often veined with black. Shropshire, Hampshire, also near Bristol. *Cost.* 139.

Purple Stone. Pale-purple, veined; composed of an angular grit, cemented by a crystalline matter, and spangled with mica. Near the sea in Flintshire. *Cost.* 140.

Bath Stone. Yellowish white; particles minute; texture uniform, compact, without glitter; soft, so as to be easily wrought with the chisel by the hand into vases, &c. Near Bath, Somersetshire.

2. Granita. *Granite.* Composed chiefly of quartz, mica, and feltspat; sometimes of garnet, basaltes, and indurated steatites. Particles visible, distinct. So hard as generally to strike fire with steel, and receive a good polish.

Black Granite, with a dark-green tinge. Frequent in the old pavement of the streets

streets of London, and on the sea shore. *Cost.* 273.

White Granite, or *Moor Stone*. White interspersed with a few large black spots and crystalline quartz. Cornwall, Devonshire, Ireland. *Cost.* 273. *Woodw. Cat.* G. e. 3.

White and green. Down, Ireland. *Cost.* 275.

Red Granite. Sometimes yellowish, generally interspersed with black mica. Devonshire. *Cost.* 276.

Yellow Granite. Interspersed with small black spots. Found in small masses on the shore near Morlin-well, in the county of Downe, Ireland. *Cost.* 280.

Green Granite. Spotted with black. On the sea-shore. *Cost.* 280.

3. Breccia. *Conglutinated Stones.* Composed of pebbles of various kinds, irregularly disposed, and cemented by various matter.

Pudding Stone. Yellowish, variegated with flints and pebbles of various colours, cemented by a jaspery substance. *Cronst.* 253. *Lin.* 12, 39.

Mill Stone. Composed of great variety of flints, pebbles, &c. cemented by a grey matter. Derbyshire. *Woodw.* tom. I. p. 29.

4. Arena.

4. Arena. *Sand.* Consisting of minute incoherent flints or stones of various colour, shape, size, and matter.

> *White.* Glittering. Found tolerably pure in several parts of England.
>
> *Yellow.* Found, near the surface, about Hampstead and Highgate, and many other parts.
>
> *Brown.* Woolwich, Black-heath, &c.

SHISTUS. *Slate.* Breaks invariably into laminæ or plates. Opaque, not flexible. Effects with steel, with acids, and with fire, various.

1. Niger. *Black Slate.* Surface hard and smooth, but not polishable. Will not strike fire with steel. When written upon, the characters are black. In some parts of England and Wales, but not frequent. *Lin.* 1, 10. *Cost.* 166.

> *Shale,* or *Bass,* or *Shiver.* Black, light, friable; rough unpolished surface. Characters, when wrote on, white. Derbyshire, in large strata up to the day, and, in other counties, generally above the coal. *Lin.* 1, 3. *Cost.* 167. *Wall. Spec.* 70.
>
> *Plate.* Black, heavy, friable, smooth, glossy. Between the laminæ are generally discovered impressions of fern and other plants. Forms a stratum immediately above the pit-coal in several

veral parts of England. *Cost.* 168. *Wall. Spec.* 67.

2. Viridis. *Green Slate.* Smooth, soft, heavy, thin, not penetrable by water. Will not strike fire with steel. Becomes purple when burnt. Characters white. In different parts of England and Wales. *Lin.* 1, 4. *Cost.* 182. Also in great abundance near Cork in Ireland.

3. Purpureus. *Purple Slate.* Smooth, but without gloss; hard, heavy, compact, composed of thin plates. Characters whitish, not penetrated by water. Sometimes slightly spangled. Will not strike fire with steel. Does not change colour in the fire. In the northern counties of England frequent. *Cost.* 175. Also in great abundance near Cork, forming huge rocks on the north side of the river; but in this variety the laminæ are neither even nor thin.

4. Ruber. *Red Slate.* Hard, heavy, strikes fire with steel. Laminæ thick and uneven. Unalterable in the fire. Near Cork.

5. Fuscus. *Brown Slate.* Pale-greenish brown, rough, coarse, hard, heavy, of a sandy texture, without brightness. Will not strike fire with steel. Becomes friable by burning, but retains its colour. Burford in Oxfordshire. *Cost.* 144.

Pale brown, or greyish, or whitish. Hard, heavy, glittering with mica; burns whitish

whitish. *Cost.* 144. 147. *Woodw. Cat.* A. b. 85. Common in the north of England.

Pale brown. Hard, heavy; texture compact but gritty, containing variety of shells, &c. Will not strike fire with steel. At Stunsfield in Oxfordshire, and at Charlwood near Bath. *Cost.* 145, 146. *Plot. Oxf.* 77. *Woodw. Cat.* G. b. 4.—10.

6. Coeruleus. *Blue Slate.* Hard, heavy, sonorous, smooth, composed of very thin plates easily separable; not penetrable by water. Characters whitish. Will not strike fire with steel. Burns brown. In many parts of England, Ireland, and Wales, particularly near Kendal in Westmoreland. Used for covering houses. *Cost.* 181. *Lin.* 1, 5. *Wall. Spec.* 66. Also at Carbery and near Kinsale in Ireland. *Smith*, vol. II. p. 373. Likewise at Denyball in Cornwall.

7. Cinereus. *Ash-coloured* or *grey Slate.* Smooth, compact, light, hard; laminæ very thin. Not penetrable by water. Will not strike fire with steel. Burns to a purplish brown. Carnarvonshire, Cumberland, Ireland, &c. *Cost.* 173.

Pale-bluish grey. Compact, coarse, hard, heavy, smooth but not glossy. Becomes friable in the air. Generally shews impressions of plants. Forms a stratum above the coal in many parts of England,

land, &c. *Cost.* 175. *Smith's Cork*, vol. II. p. 373.

Dead ash colour. Light, rough, coarse, friable, without the least gloss, of a sandy texture, composed of very thin laminæ; burns whitish. In several parts of England. *Cost.* 144.

8. Lapis Hibernicus. *Irish Slate.* Lead colour; writes on paper like black-lead. Dr. Hill says he extracted a considerable proportion of alum from this slate. If it had not been asserted by a person of the Doctor's known integrity, I would not believe a word of it.

FLUOR. Appearance sparry or crystalline. Will not strike fire with steel. Does not ferment with acids. Vitrescent *per se*, tho' sometimes refractory, but easily fusible with borax, or calcareous or other earths. When gradually heated shines like phosphorus; but its light vanishes before it becomes red-hot. Promotes the fusion of ores. *Cronst.* 108.

1. Crystallizatus. *Crystallized Fluor.* White, or blue, or purple, or green, or red. In iron and copper mines.

Sparry, but less regular than spar.

Irregular.

Cubic.

QUARTZUM.

QUARTZUM. Does not effervesce with acids. Strikes fire with steel. Does not after burning fall to powder either in the air or in water. When pure cannot be melted *per se*, but with alkaline salts most readily into glass. Texture solid, uniform; particles homogeneous, invisible, impalpable. Breaks in various directions.

1. Crystallus. *Crystal.* Diaphanous, naturally hexagonal, columnar, pyramidal at one or both ends, but frequently rounded. Luid. 1.

   Colourless. Found near West Carbery. Smith's *Cork*, vol. II. p. 382. *Woodw.* tom. I. p. 31. 158.

   Milky. Woodw. tom. I. p. 158.

   Red. Woodw. tom. I. p. 160.

   Brown. Woodw. tom. I. p. 160.

   Yellow. Woodw. tom. I. p. 160.

   Purple. Woodw. tom. I. p. 160.

   Black. Woodw. tom. I. p. 160.

2. Silex. *Flint.* Semipellucid, detached, often somewhat spherical; breaks in convex and concave polished fragments. Decays when exposed to the air, and commonly surrounded by a rough crust.

   Common

Common Flint. Colour uniform, not ſtreaked or veined. Dark or light horn-colour or yellowiſh. Frequent in chalk-pits, gravel-pits, and on the ſea-ſhore. *Cronſt.* 67. *Lin.* II, 1.

Onyx. The hardeſt of this ſpecies, confiſting of zones of different colours. *Woodw.* tom. I. p. 36. c. 192.

Agate. Clouded, or veined, or ſpotted, with different colours. *Woodw.* tom. I. p. 32.

3. Jaſpis. *Jaſper.* Opaque. Texture when broken like dry clay. Fragments not convex and concave. Leſs hard than flint, and melts more readily into glaſs. Does not decay in the air. Takes a fine poliſh. *Cronſt.* 69. *Lin.* II, 14.

Sparry or *cryſtalliſed*, often rhomboidal. Generally mixt with other foſſiles in the compoſition of granites, porphyry, &c. *Cronſt.* 72. *Lin.* 5, 12. *Wall.* 125.

4. Granatus. *Garnet.* Diaphanous, dark-bluiſh red, ſpherical, with an indeterminate number of facets, or ſides, containing a ſmall portion of iron or tin. Frequent, though ſmall, in micaceous ſtones, particularly in Scotland. *Coſt.* on *Cronſt.* 84. *Wall.* vol. I. 223.

C 2   5. Baſaltes.

5. Basaltes. *Shirl.* Opaque, ponderous, generally black or green, glossy, crystallized, prismatical, of an indeterminate number of angles. Forming an immense rock called the Giants Causeway in Ireland, and in small pieces in the tin-mines in Cornwall. *Cronst.* 80. *Wall.* vol. I. p. 261. *Cost.* 252.

CLASS II.

SALTS.

Sapid, soluble in water, and recoverable by evaporation in their pristine angular form; not inflammable, nor ductile.

SAL ACIDUM. When most pure, fluid, and mixt with a considerable proportion of water. Tastes sour; changes blue vegetable juices red; dissolves calcareous earth with effervescence; unites violently with alkalis, with which it has the greatest affinity, except with phlogiston. A stronger attraction to zinc than to any other metal.

1. Vitriolicum. *Vitriolic Acid.* When mixt with the least possible quantity of water, its specific gravity is to water as 18 to 10. Dissolves silver, tin, antimony, mercury, zinc, iron, copper, lead. United with calcareous earth forms gypsum; with argillaceous earth, alum; with phlogiston, sulphur; with metals, vitriol; with alkaline salts, neutrals. Precipitates all solutions in the nitrous and vegetable acids. With spirit of wine produces æther; with water generates heat; with

water and steel filings takes fire. Exists in native Glauber's salt; in metallic salts whose basis is iron, copper or zinc; in alum, sulphur, in various waters, and in the atmosphere, but most frequently obtained from sulphur or vitriol. *Pharm. Med.* 2d edit. p. 5. *Cronst.* 129.

2. Muriaticum. *Acid of Sea-Salt.* When most concentrated is of a yellow colour. Specific gravity to water as 12 to 10. Dissolves tin, lead, iron, copper, zinc, antimony. Mixt with the nitrous acid forms *aqua regia*, which dissolves gold. With alkaline salts forms various neutrals; with calcareous earth *sal ammoniacum fixum*. Generally obtained from common *salt*. May be disunited from alkaline salts by the vitriolic acid. *Pharm. Med.* p. 9.

SAL ALKALI. Tastes acrid, urinous. Unites with acids with violent effervescence. Changes blue vegetable juices green. With oil forms soap.

1. Fixum. *Fixed mineral Alkali*; *Natron* of the antients. Shoots readily into prismatical crystals. Dissolves every species of quartz forming glass. Falls to powder in the air. Promotes the fusion of metals, and precipitates them when dissolved in acids. With acids forms various neutrals, which see. Obtained from sea salt, mineral waters, marine plants, and in considerable quantity from a white efflorescence on walls

not

2. Volatile. *Volatile mineral Alkali.* Differs in no respect from that obtained by distillation from animal and vegetable substances. Contained in most clays. *Cronst.* 148.

SAL NEUTRUM. Composed of an acid and an alkaline salt; therefore will not effervesce with either. Form regular crystalline.

1. Sal Glauberi. *Glauber's Salt.* Vitriolic acid and fossile alkali. Crystals hexagonal. Soluble in an equal weight of water. Found in some mineral waters, but generally the produce of art. Fusible in a moderate degree of heat. *Pharm. Med.* p. 15.

2. Sal Commune. *Common Salt,* or *Sea Salt.* Muriatic acid and fixt fossile alkali. Crystals cubic. Obtained from sea-water and from salt-springs in Cheshire, Worcestershire, Hampshire, Staffordshire, Northumberland. *Pharm. Med.* p. 21. *Cronst.* p. 139. *Lin.* 16. 1. 3. *Woodw.* tom. I. p. 170.

3. Sal Ammoniacum commune. *Common Sal Ammoniac.* Muriatic acid and volatile alkali. Crystals indeterminate. Mixt with common salt generates extream cold. Liquifies in the air. Soluble in vinou spirits.

spirits. Soluble in three times its own weight of water. Obtained by sublimation from every species of soot. *Pharm. Med.* p. 22.

SAL METALLICUM. *Metallic Salt.* Metal diſſolved in the vitriolic acid; cryſtallized. Soluble in ſixteen times its weight of water.

1. Vitriolum Cœruleum. *Blue Vitriol.* Vitriolic acid and copper. Cryſtals depreſſed, rhomboidal, with twelve ſides. Taſte aſtringent, acrid, diſagreeable. Fuſible, and in a ſtrong fire calcinable. Generally the produce of art, but ſometimes native, diſſolved in Ziment waters near copper mines, particularly in the county of Wicklow in Ireland. *Pharm. Med.* p. 26. *Lin.* 18, 3. *Cronſt.* 131.

2. Vitriolum Viride. *Green Vitriol.* Vitriolic acid and iron. Cryſtals rhomboidal, ſhort, thick, pale-green, pellucid. Taſte aſtringent, ſweetiſh. Eaſily ſoluble in water, and calcinable in the fire. Generally artificial, but often diſſolved in mineral waters. *Pharm. Med.* p. 27. *Cronſt.* p. 130. *Lin.* 18, 1. *Woodw.* tom. 1. p. 171.

3. Vitriolum Album. *White Vitriol.* Vitriolic acid and zinc. Cryſtals, according to Linnæus, priſmatical, dodecaedral. Taſte aſtringent, ſweetiſh. In mines, often

often mixt with copper and iron. *Cronſt.* 131. *Pharm. Med.* p. 28.

4. Arſenicum. *Arſenic.* White, volatile in the fire. Cryſtals priſmatical, octoedral. Soluble by boiling in water. Eaſily cryſtallized by ſublimation with phlogiſton, and ſometimes found native in a cryſtalline form. See *Arſenicum* as a metal.

SAL TERREUM. Vitriolic acid and earth, ſeparable by the interpoſition of an alkali.

1. Alumen. *Rock Alum.* Vitriolic acid and argillaceous earth. Cryſtals octoedral, pyramidal. Liquifies in the fire. Taſte exceſſively aſtringent. Soluble in fourteen times its weight in water. Obtained by art from ſtones of different kinds near Whitby in Yorkſhire. *Cronſt.* 133. *Pharm. Med.* p. 30. *Lin.* 17. *Woodw.* tom. I. p. 170.

Feather Alum. Reſembles white feathers. In ſmall quantities on decayed alum-ſtones. *Cronſt.* 133. *Lin.* 17. 1.

2. Sal Ammoniacum Fixum. *Fixt Sal Ammoniac.* Muriatic acid and calcareous earth. In ſea-water, and at the bottom of the pans at the ſalt-works. *Cronſt.* 27.

3. Sal Catharticum Amarum. *Bitter purging Salt,* or *Magneſia, Glauber's Salt,* or *Epſom Salt.* Vitriolic acid and magneſia alba. Cryſtals hexagonal,

hexagonal, prismatical. May be decomposed by an alkali. In the Epsom water, and in many other springs. *Pharm. Med.* p. 30. *Wall.* 339.

4. Selenites. Vitriolic acid and calcareous earth. Crystallizes in thin laminæ. Decomposed by an alkali. Very frequent, in small proportion, in mineral and in common spring water. See *Gypsum*, Numb. 5.

CLASS

CLASS III.

INFLAMMABLES.

Soluble in oil, but not in water; readily take fire, and are in a great measure consumed.

BITUMEN. Mineral phlogiston united with mineral acid and other fossile substances.

1. Petroleum. *Fossile* or *Rock Oil*. Highly inflammable, fluid, pellucid, fragrant, light, pale-brown, but exposed to the air becomes thick and black. Miscible with essential oils, but not with vinous spirits. Found floating on the water of certain springs in Persia, Italy, and in England, &c. particularly at Wenlock and Pitchford in Shropshire, Wigan in Lancashire, Libeston in Midlothian, Scotland. *Sibald. Prodr.* part II. l. 4. c. 4. *Cronf.* 152. *Pharm. Med.* p. 35. *Lin.* 21, 1, 2.

2. Electrum. *Amber*. Phlogiston united with muriatic acid. Solid, hard, brittle, light; texture compact, uniform; particles homogeneous, invisible. Takes a fine polish; fragrant when rubbed, and highly elec-

electrical. Found in the earth, and on the sea-shore, in nodules from half an inch to two inches in diameter.

Pale-yellow, perfectly pellucid. On the sea-shore of Yorkshire, Norfolk, &c. also in clay-pits near London.

Brown, transparent. In clay-pits in Leicestershire. *Woodw.* tom. I. p. 168.

3. Gagas. *Jet.* Wood impregnated with the inflammable substance of pit-coal. Black, solid, dry, opaque, light, capable of a fine polish, electrical. Texture evidently that of wood. Burns slowly with a white flame. Found in detached pieces in strata of earth and stone in the northern counties, and in clay-pits near London. *Hill. Foss.* 413. *Cronst.* 263. *Lin.* 2;. 8. *Brand. fos. Lant. fig.* 121. *Woodw.* tom. I. p. 167.

4. Lithantrax. *Coal.* Phlogiston united with argillaceous earth and vitriolic acid. Black, solid, opaque, dry, brittle, glossy. Found in large strata, splitting nearly in a horizontal direction.

Pit Coal. Friable. Newcastle, &c.

Stone Coal. Hard, heavy, brittle; splits in the fire, burns briskly, and flames much. In the north of England, &c.

Cannel Coal. Bright, light, splits in any direction, does not colour the hands,

and

and takes a good polish. *Wood.* tom. I. p. 165.

Kilkenny Coal. Lights flowly, and burns almoſt entirely without ſmoke. Seems not to differ in the leaſt from the coal of Pembrokeſhire in Wales, when that is ſeparated from the coal-duſt with which it is mixt. Near Kilkenny in Ireland.

5. Turfa. *Turf.* Mould impregnated with bitumen, interwoven with roots of vegetables. Cut in the ſhape of bricks, and uſed for fuel in many parts of this kingdom. *Cronſt.* 264. *Lin.* 54, 6. *Wall.* 356.

SULPHUR. Phlogiſton united with a large proportion of vitriolic acid. Yellow; fuſible in a moderate degree of heat; burns with a blue flame; totally volatile in the fire. Not found native, or pure, in theſe kingdoms.

1. Pyrites. *Mundic.* Sulphur combined with iron. Pale-yellow, of a metallic appearance, not cryſtallized. Strikes fire with ſteel. In various mines. *Cronſt.* 155. *Lin.* 22, 5. *Wall.* 379.

2. Marcaſita. *Marcaſite.* Sulphur combined with iron, copper, &c. cryſtallized. Strikes fire with ſteel. Yellow, or white, or grey. *Cronſt.* 155. *Lin.* 22, 3. *Wall.* 384.

3. Molyb-

3. **Molybdæna.** *Black-lead.* Sulphur combined with iron and tin. Used for pencils, &c. Barrowdale, near Keswick in Cumberland. *Cronst.* 156. *Lin.* 25, 1, *Woodw.* tom. I. p. 185.

Sulphur is likewise found combined with other metals and semimetals; which see.

CLASS

CLASS IV.

METALS.

Of all mineral bodies, most ponderous, Fusible, but resuming their original properties and a convex surface when cold; even after calcination, by the addition of phlogiston.

I. METALS, *properly so called*.

AURUM. *Gold.* Yellow. The heaviest metal (viz. to water as 19640 to 1000.) and most ductile. Unalterable by fire. Soluble only in *aqua regia* and *hepar sulphuris*. Amalgamates readily with quicksilver.

1. Nativum. *Gold-dust* or *sand*. Said to be found in small quantity, in some rivers in Scotland. *Cronst.* 167. *Lin.* 35, 1.

ARGENTUM. *Silver.* White. Specific gravity to water as 11,091 to 1000. Most ductile, except gold. Sonorous. Soluble in the nitrous and vitriolic acids, not in *aqua regia*. Unalterable in the fire. Easily amalgamates with quicksilver.

1. Minerali-

1. Mineralizatum. Mineralized with sulphur and other metals. Frequent in most of our lead and copper ores, but in no great proportion.

STANNUM. *Tin.* White. Most easily fusible, but least ductile, and lightest of all metals. Soluble in *aqua regia*, vitriolic and muriatic acids. Unites with all metals and semimetals, rendering them sonorous and brittle. Amalgamates easily with quicksilver. Specific gravity to water as 7,500 to 1000. Cornwall.

1. Nativum. *Native Tin.* Very rare. The existence of native tin is generally denied; but the Royal Society is in possession of an undoubted specimen, sent from Cornwall.

2. Crystallizatum. *Tin Crystals* or *Tin Grains.* Mineralized by an admixture of arsenical earth. Opaque, spherical, polygonal, glossy, heavy, yellowish or brown, or black. *Cronst.* 181. *Lin.* 30, 1. *Wall.* 548. *Woodw.* tom. I. p. 201. tom. II. p. 30.

3. Amorphum. *Tin-stone.* Combined with an arsenical earth and some iron. Blackish-brown or yellowish, resembling a common stone, opaque and very heavy. On the coast near Penrose in Cornwall. *Cronst.* 181. *Lin.* 30, 3. *Wall.* 550. *Woodw.* tom. I. p. 202.

4. Molyb-

4. Molybdæna. *Black-lead.* See Sulphur.

5. Granatus. *Garnet.* See Quartzum.

PLUMBUM. *Lead.* Bluish white when first cut or broken. Specific gravity to water as 11,350 to 1000. Less hard, less elastic, less tenacious, less sonorous, than any other metal. Soluble in all acids and alkaline solutions. Fusible before ignition, and easily calcined.

1. Galena. *Lead Glance.* Mineralized with sulphur and a little silver. Opaque, bluish, glossy, composed of large or small cubes. Contains a very large proportion of lead. In various parts of England. *Cronst.* 186. *Lin.* 31, 3. *Wall.* 529. *Smith's Cork*, vol. II. p. 393. *Woodw.* tom. I. p. 211. tom. II. p. 27.

2. Stibiatum. *Antimonial Lead-ore,* or *Lead-trail.* Mineralized with sulphur, antimony, and silver. Opaque, radiated, shining. *Lin.* 31, 5. *Woodw.* tom. I. p. 211.

3. Crystallinum. *Lead Crystals.* Diaphanous, prismatical or pyramidal, of no determinate number of sides, but most frequently hexagonal; or white, or yellowish, or greenish. Mendip, Somersetshire. *Cronst.* 186. *Lin.* 31, 2. *Woodw.* tom. I. p. 214. N°. 58, 59, 132.

4. Spatosum. *Lead Spar.* White or grey, often yellowish, without the least metallic appearance. Near Keswick, in Cumberland; also near Bristol. *Wall.* 535. *Woodw.* tom. I. p. 214.

5. Calciforme. *Lead Ochre,* or *Native Cerussa.* A white powder sometimes found on the surface of lead-glance. *Cronst.* 184.

CUPRUM. *Copper.* Yellowish-red. Specific gravity to water as 9 to 1. Most ductile, except gold and silver. Most elastic, except iron. Most sonorous. Soluble in all acids and alkaline solutions, oils and water. Requires almost as great a degree of heat as iron before it melts.

1. Nativum. *Native* or *Virgin Copper.* Solid, malleable; branched or fibrous, or foliated; generally adhering to other fossile substances. Cornwall, Isle of Man, &c. *Cronst.* 190. *Lin.* 33, 2. *Wall.* 499. *Woodw.* tom. I. p. 195, 197. tom. II. p. 24.

Ziment Copper. Granulated, friable. Precipitated from the vitriolic acid by the immersion of iron or otherwise. In the county of Wicklow, in Ireland. *Cronst.* 190. *Lin.* 33, 1. *Wall.* 501.

2. Coeruleum montanum. *Mountain-blue.* Copper united with calcareous earth, dissolved

folved and precipitated from an alkaline menſtruum. Soluble in *aqua fortis* with effervefcence. In the mines in Derbyſhire, &c. *Coſt.* 104. *Cronſt.* 36, 190. *Lin.* 30, 4. *Woodw.* tom. I. p. 195.

3. Viride montanum. *Mountain-green.* Copper united with earth, diſſolved by an acid. In the copper mines of England and Ireland, and the Iſle of Man. *Coſt.* 106. *Cronſt.* 190. *Lin.* 50, 3. *Woodw.* tom. I. p. 197.

4. Rubrum. *Glaſs Copper Ore.* Hard, brittle, red or purple, or brown. Generally found with native copper. *Cronſt.* 191. *Lin.* 33, 9. *Woodw.* tom. I. p. 196.

5. Cinerum. *Grey Copper Ore.* Mineralized with ſulphur alone. Solid or diced. Soft ſo as to cut with a knife. *Cronſt.* 192. *Lin.* 33, 6. *Wall.* 510.

6. Pyrites. *Copper Pyrites*, or *Mundic.* Yellow or yellowiſh, or ſometimes greeniſh. Copper mineralized with iron, and frequently a ſmall proportion of arſenic; marcaſitical. Cornwall, &c. *Cronſt.* 193. *Lin.* 33, 4, 5, 7. *Wall.* 514. *Smith's Cork*, vol. II. p. 386. *Woodw.* tom. I. p. 179, 193, 198.

7. Albidum. *White Copper Pyrites.* Mineralized with ſulphur, iron, and a conſiderable proportion of arſenic. Texture compact.

Yellowish-white. Heavy, rich, but scarce. *Cronst.* 194. *Lin.* 32. 8.

FERRUM. *Iron.* Attracted by the magnet; most elastic; most sonorous, except copper. Soluble in all acids and alkaline solutions, water and air. Most difficult of fusion; least malleable. Specific gravity to water as 8 to 1.

1. Ochra. *Iron Ochre.* Yellow, or brown. Iron which has been dissolved in the vitriolic acid, and precipitated in form of a loose powder or a friable mass. In the fissures of iron mines in Dean Forest, &c. *Woodw.* tom. II. p. 1. *Cronst.* 196. *Lin.* 50, 1. 49, 5. *Wall.* 478. *Smith's Cork*, vol. II. p. 368. *Woodw.* tom. I. p. 8. 230. N°. 62.

2. Hæmatites. *Bloodstone.* Hard, heavy, red, or brown, or grey; yielding a red powder, when rubbed. Contains a large proportion of iron, tho' not attracted by the loadstone. Generally found in masses, convex on one side, and of a fibrous texture. Near Whitehaven, in Cumberland, &c. *Woodw.* tom. I. p. 228. *Cronst.* 197. *Wall.* 469. *Lin.* 32, 2, 22, 23.

3. Crystallinum. *Crystalline Ore.* Composed of small shining cubical or octoedral particles, brown, resembling marcasite. Not malleable, nor attracted by the loadstone.

stone. *Wall. Spec.* 252. *Cronst.* 197. *Lin.* 32, 2, 2. Forest of Dean, Langron in Cumberland. *Woodw.* tom. I. p. 229. N° 60, 61. tom. I. p. 225, n. 16.

4. Selectum. *Common Iron Ore.* Solid, brown, or blackish. Yields a black powder when rubbed; is attracted by the magnet. *Cronst.* 203. *Wall. Spec.* 254. *Lin.* 32, 4, 8. *Smith's Cork*, vol. II. p. 390. *Woodw.* tom. I. p. 227.

5. Cœrulescens. *Bluish Ore.* Generally brown on the outside. Hard and heavy; solid or scaly; attracted by the magnet; rich in iron, and easily melted. *Lin.* 32, 19. *Wall. Spec.* 256.

6. Magnes. *Magnet*, or *Loadstone*. Mineralized by a small proportion of sulphur. Blackish, or grey, or brown. Solid or granulated. Attracts iron, and points north and south. *Cronst.* 202. *Lin.* 32, 27. *Wall. Spec.* 259. Mendip Hills, Somersetshire; Devonshire. *Woodw.* tom. I. p. 234. tom. II. p. 21.

7. **Micaceum.** *Glimmer*, or *Eisenman*. Mineralized with sulphur. Refractory. Scaly, shining, brittle, dark grey. *Lin.* 32, 18. *Cronst.* 203.

D 3   8. Mag-

8. Magnefia. *Manganefe.* Blackifh, radiated, fhining; radii generally converging to a centre. Sometimes fcaly. Friable, colouring the fingers. Not attracted by the magnet. Gives a violet-colour to glafs. Mendip, Somerfetfhire. *Woodw.* tom. I. p. 234. *Cronf.* 121. *Lin.* 25, 2. *Wall. Spec.* 264.

Iron is alfo found in the ores of other metals, and in mineral waters.

II. SEMI-METALS.

Not malleable.

VISMUTUM. *Bifmuth,* or *Tin-glafs.* Yellowifh-white; foft, yet brittle; texture laminated. Specific gravity to water as 9,700 to 1000. Eafily fufible; volatile in the fire. Soluble in the mineral acids, and in *aqua regia.* Amalgamates eafily with quickfilver, and unites it fo intimately with other metals, efpecially lead, as to carry them through leather without feparation.

1. Nativum. *Native Bifmuth.* Solid in fmall cubes, or fuperficial like a cruft on other bodies. Melts in the flame of a candle. Found with tin, cobalt and copper ores.
Cronft.

Cronft. 211. *Lin.* 28, 1. *Wall. Spec.* 243.

2. Ochra. *Flowers of Bismuth.* Generally yellowish, in form of an efflorescence on various ores, &c. *Cronst.* 211. *Lin.* 50, 7. *Wall. Spec.* 245.

3. Mineralizatum. *Bismuth Ore.* Mineralized with sulphur. Grey, of a radiated appearance, composed of thin square laminæ, resembling lead-glance. *Cronst.* 212. *Lin.* 28, 2, 4. *Wall. Spec.* 244.

ZINCUM. *Zinc,* or *Spelter.* Bluish-white. Texture fibrous, or of parts resembling flat pyramids. Specific gravity to water as 7 to 1. Melts easily, and burns with a yellowish-green flame, subliming in white smoke. Unites with all metals, except bismuth, rendering them volatile. Soluble in all acids. Its filings are attracted by the magnet. Separable from copper by mercury.

1. Cryftallinum. *Crystallized Zinc.* A pure calx of zinc. Grey, resembling lead spar, or an artificial glass of zinc. *Cronst.* 215. *Lin.* 27, 1.

2. Calaminaris. *Calamine.* A calx or earth of zinc, mixt with iron. Yellowish, or brown. Mendip, Somersetshire. *Cronst.* 216. *Lin.* 27, 5. *Wall. Spec.* 248.

248. *Woodw.* tom. I. p. 184. tom. II. p. 19.

3. Rapax. *Blend.* Mixt with sulphur and iron. Yellow or black, or brown or red; scaly, often semi-diaphanous, resembling glass. *Cronst.* 217. *Lin.* 27, 6, 8. *Wall. Spec.* 249.

4. Mineralizatum. *Ore of Zinc.* Mineralized with sulphur and iron. Of a metallic appearance; bluish-grey, scaly, or cubical. *Cronst.* 216. *Lin.* 27, 3. *Wall. Spec.* 247.

ANTIMONIUM. *Antimony.* White, like silver. Texture fibrous. Brittle. Specific gravity to water as 7,500 to 1000. Volatile in the fire. Volatilizing other metals, except gold. Soluble in spirit of salt, and in *aqua regia*, from the last of which it may be precipitated by water. If previously melted with lime, it will amalgamate with mercury. Prevents iron from being attracted by the magnet.

1. Striatum. *Striated Antimonial Ore.* Mineralized with sulphur. Bluish-grey, shining, striated, and scaly. Melts in the flame of a candle. *Cronst.* 222. *Lin.* 26, 3. *Wall. Spec.* 238. Cornwall, Flintshire. *Woodw.* tom. I. p. 184. tom. II. p. 20.

2. Crystal-

SEMI-METALS.

2. Cryſtallizatum. *Cryſtallized Antimony.* Striated within, but externally of a cryſtalline appearance. *Wall. Spec.* 241. *Cronſt.* 222. *Lin.* 26, 2. *Woodw. tom.* I. p. 184. Cornwall.

3. Rubrum. *Red Antimonial Ore.* Mineralized with ſulphur and arſenic. Alſo ſtriated; but its fibres finer than either of the preceding ſpecies. *Cronſt.* 223. *Lin.* 26, 4. *Wall. Spec.* 242.

ARSENICUM. *Arſenic.* Originally the colour of lead; but being expoſed to the air turns yellow, and then black. Texture laminated. Extremely volatile in the fire, riſing in white ſmoke, and ſmelling like garlick. Yields a regulus on being melted with potaſhes and ſoap; or by ſublimation, mixt with phlogiſton. Soluble in acids, and even in water by boiling. Unites with all metals. Specific gravity of this regulus to water as 8,308 to 1000.

1. Nativum. *Native Arſenic.* Soft as black lead, compoſed of hemiſpherical laminæ. *Cronſ.* 226. *Lin.* 23, 1.

2. Calciforme. *White Arſenic.* In form of a white powder, ſometimes cryſtalline. *Cronſt.* 226.

3. Auripigmentum. *Orpiment.* Mixt with ſulphur. Yellow or greeniſh, or red; foliated,

liated, shining. *Cronst.* 227. *Lin.* 22; 2, 4.

4. Mineralizatum. *White Mundic,* or *White Pyrites,* or *Marcasite.* Mineralized with sulphur and iron. Texture irregular, or cubical, or prismatical. *Cronst.* 228. *Lin.* 23, 5, 6.

Arsenic is also discovered in tin grains, lead spar, cobalt ore, copper ore, antimony, &c.

COBALTUM. *Cobalt.* Whitish-grey. Hard and brittle, not shining. Fixt in the fire. Its calx tinges glass deep blue. Soluble in the vitriolic and nitrous acids, and in *aqua regia,* tinging them red. Will not unite with bismuth alone; nor amalgamate with quicksilver. Specific gravity to water as 6 to 1.

1. Calciforme. *Black Cobalt.* Mixt with iron without arsenic. Either friable in form of an ochre, or a slag hard and glossy. *Cronst.* 231. *Lin.* 29, 4.

2. Ochra. *Cobalt Ochre,* or *Cobalt Flowers.* Mixt with the calx of arsenic. Pale red, or yellowish. An efflorescence on cobalt ores. *Cronst.* 232. *Lin.* 50, 8. *Wall. Spec.* 235.

3. Arsenicale. *Cobalt Ore.* Mineralized with arsenic and iron. Solid, resembling steel, or crystallized. *Cronst.* 232. *Lin.* 29, 2. *Wall. Spec.* 231.

4. Crystal-

4. Cryſtallinum. *Cryſtalline Cobalt Ore.* Mineralized with ſulphur, iron, and arſenic. Reſembles the laſt ſpecies, but of a lighter colour. *Cronſt.* 234. *Lin.* 29, 1. *Wall. Spec.* 234.

5. Niccolum. *Copper-nickel.* Mineralized with ſulphur, arſenic, and iron. Reddiſh-yellow, reſembling copper. *Cronſt.* 238. *Lin.* 33, 19. *Wall. Spec.* 229.

CLASS

CLASS V.

PETREFACTIONS.

Animals, or parts of animals or vegetables, changed into a fossile substance.

HELMINTHOLITHUS. *Vermes.* Class VI. vol. I. *Lin. Gen.* 41.

1. Ammonita, or Cornu Ammonis, or Nautilus. *Serpent Stone.* Flat, spiral, representing a worm or small serpent, coiled up; of various dimensions, and variously striated; ridged and studded. Found frequently in strata of earth and stones; also on the sea-shore. Whitby, in Yorkshire; Pyrton Passage, in Glocestershire; Stoke, in Somersetshire. *Wall. Spec.* 374, 387. *Woodw.* tom. I. part ii. p. 24.

2. Anomites. .... Bivalve; one valve gibbous, and often perforated at the base; the other plane, and less. Hinge without teeth. Found in great abundance in various parts of England; particularly at Sherborne, in Glocestershire. Vol. I. p. 200. *Wall. Spec.* 397. *Luid.* cap. 14.

14. *Woodw.* tom. I. part ii. p. 45 to 51.

3. Gryphites. .... Bivalve, oblong, somewhat resembling a boat, but narrow, and remarkably curved upwards at one end; the other valve plane. Vol. I. p. 200. *Wall. Spec.* 396. *Luid.* cap. 9. *Lift. Angl.* lib. iv. fol. 45. In chalk-hills, &c.

4. Judaicus. .... Supposed to be spines of the Echinus; resembling an olive or small cucumber, with a short stem; smooth, or striated, or studded. Found in many parts of England; particularly in the chalk-pits in Kent. *Wall. Spec.* 400. *Luid.* cap. 16. *Plot Oxf.* tab. 6.

5. Echinites. .... Roundish, resembling a button; frequently flinty, sparry, or cretaceous; with tubercles and lines regularly diverging from the center, with marks of an aperture above and beneath. Surry, Eflex, Kent, Middlesex. Vol. I. p. 195. *Wall. Spec.* 399. *Luid.* cap. 15. *Woodw.* tom. I. part ii. p. 64. In chalk and gravel pits.

6. Aftrion. *Sea Star.* Minute, reddish-white, in form of a star or wheel, with four or five radii; somewhat convex in the center. Detached joints of the next species. In chalk-pits, &c. Vol. I. p. 195. *Plot Oxf.* 85. n. 16. *Wall. Spec.* 356.

7. Afteria.

7. Asteria columnaris. .... Cylindrical, but pentangular; generally about an inch long, and the thickness of a quill; often crooked, with a star of five radii at each extremity. In various parts of England and Wales. Near Marston-truffel, in Northamptonshire; near Whitton, in Lincolnshire; Skerborne, in **Glocester-shire**; Shughborough, in Warwickshire; PyrtonPassage, &c. *Lin. Syst. Nat.* p. 1288. n. 5. *Wall. Spec.* 359. *Luid* cap. 17. *Plot Oxf.* tom. II. fol. 2, 3. *Morton Northamp.* p. 239. *Woodw.* tom. I. part II. p. 80.

8. Entrochus. .... Cylindrical, generally about an inch long, sometimes much less, and sometimes three inches; composed of several flat, round joints, with radii on each disk, and perforated through the middle. Stainton, in Cumberland; in the river near Moreland, in Westmoreland; Mendip, Somersetshire; Stone Quarry, near Mask, in Yorkshire; King's Weston, Glocestershire. *Luid.* N°. 1133. *Wall. Spec.* 357. *Syst. Nat.* p. 1288. *Woodw.* tom. I. part II. p. 78.

9. Belemnites. .... Cylindrical, but conical at one, sometimes at both ends; smooth; generally about the length and thickness of a finger, with a conical cavity at the base, which cavity is often filled with a nucleus, called alveolus. When broken, appears to be composed of longitudinal fibres, with others from the centre to the circum-

circumference. In various strata, particularly lime-stone, in many parts of England and Wales. *Syst. Nat.* p. 1295. and vol. III. p. 170. *Wall. Spec.* 355. *Luid.* cap. 23. *Woodw.* tom. I. p. 106.

10. Tubiporus. .... A congeries of coralline tubes, parallel or variously curved ; found frequently loose in different strata, and often immersed in stone. On the shore near Sunderland, in the bishopric of Durham ; Yorkshire, &c. *Syst. Nat. Gen.* 336. *Luid.* cap. 17. *Wall. Spec.* 330. *Woodw.* tom. I. p. 130, 132. tom. II. p. 10.

11. Madreporus. Coral, branched, with stars at the extremity of each branch. Broadwell Grove, Glocestershire, &c. *Syst. Nat.* p. 1272. *Wall. Spec.* 328. *Luid.* cap. 2. *Woodw.* tom. I. p. 131.

12. Millepora. Coral, branched, with the surface and extremities punctured, as if pierced with the point of a needle. Near Dudley, Staffordshire, &c. *Syst. Nat.* p. 1282. *Wall. Spec.* 329. *Luid.* cap. 2. Vol. I. p. 208. *Woodw.* tom. I. p. 130.

13. Astroites. *Star Stone.* Coral, solid, texture tubular, of various shape, often resembling a mushroom ; surface covered with stars, which are the extremities of the tubes of which it is composed. Glocestershire, Northamptonshire, &c. In gravel-

gravel-pits. Near Glanmire river, Cork, *Smith*, vol. II. p. 381. *Luid.* n. 160. *Woodw.* tom. I. p. 142.

14. Trochus. *Top-Shell.* Single, spiral, subconic; apertures somewhat angular, or oval; columella oblique. *Syst. Nat. Gen.* 326. Vol. I. of this work, p. 202. *Brand. fof. hant.* fig. 1, 2, 3, 4, 5, 6. *Luid.* cap. 7. *Wall. Spec.* 377. *Woodw.* tom. I. part ii. p. 31.

15. Turbo. *Screw-Shell.* Single, spiral, solid; aperture small, orbicular, contracted, entire. Vol. I. 204. *Syst. Nat. Gen.* 357. *Brand. fof. hant.* fig. 7, 8, 27, 47, 48, 49, 50. *Luid.* p. 20.

16. Dentalium. *Tooth-Shell.* Single, tubular, tapering to a point, straight or nearly so, pierced at each end. *Syst. Nat.* p. 1263. *Wall. Spec.* 373. *Brand. fof. hant.* fig. 9, 10, 11. *Woodw.* tom. I part ii. p. 23.

17. Serpula. .... Single, tubular, cylindrical, almost straight, smooth. *Syst. Nat.* p. 1264. *Brand. fof. hant.* fig. 12. *Woodw.* tom. I. part ii. p. 37.

18. Murex. .... Single, spiral; aperture oblong, ending in a long straight beak or canula. *Syst. Nat.* p. 1213. *Brand. f.f. hant.* fig. 13, 17, &c. p. 2.

19. Buccinum. .... Single, spiral; the first volution much larger than the rest; aperture

aperture oblong, ending in a short dexter canula. *Syst. Nat.* p. 1196. *Brand. fof. hant.* fig. 14, 15, 16, 18, 19, 20, 43, 56, 63, 71. *Woodw.* tom. I. part ii. p. 36, 78.

20. Conus. ... Single, convoluted, turbinated; aperture long, narrow, plain, base entire; columella smooth. *Syst. Nat.* p. 1165. *Brand. fof. hant.* fig. 21, 22, 24.

21. Voluta. ..... Single, spiral, sub-cylindrical; aperture long.

22. Bulla. ..... Single, convoluted, oblong or oval, smooth; aperture long, narrow; base entire; columella oblique. *Syst. Nat.* p. 1181. *Brand. fof. hant.* fig. 29, 61, 75.

23. Strombus. .... Single, spiral; aperture with the lip generally dilated, ending in a sinister canula. *Syst. Nat.* p. 1207. *Brand fof. hant.* fig. 42, 64 to 69, 76. Columella often plicated and spinous.

24. Helix. *Snail.* Single, spiral, thin, brittle; aperture contracted, lunated on the inside, subrotund. Vol. I. p. 204. *Syst. Nat.* p. 1241. *Brand. fof. hant.* fig. 57, 58, 59, 60. *Woodw.* tom. I. part ii. p. 108.

25. Ostrea. *Oister,* and *Scallop.* Bivalve, inequivalve. Hinge without teeth, with an oval cavity. *Vulva anusve nulla.* Trans-

Vol. III.   E   verse

verse striæ. Vol. I. p. 200. *Syst. Nat.* p. 1144. *Brand. fos. hant.* fig. 83, 88, 107. *Luid.* cap. 8, &c. *Woodw.* tom. I. part ii. p. 38, 42. tom. II. p. 40. Glocestershire; Berkshire near Reading; Woolwich, Kent; Oxfordshire; Northamptonshire.

26. Chama. Bivalve, thick; hinge a gibbous callus inserted in an oblique groove. *Lin. Syst. Nat.* p. 1137. *Brand.* fig. 84, 85, 86, 87, 100.

27. Tellina. Bivalve. In the fore part of one shell a convex, in the other a concave fold. Hinge of three teeth. Glocestershire. Vol. I. p. 198. *Syst. Nat.* p. 1116. *Brand. fos. hant.* fig. 89, 102. *Woodw.* tom. I. part ii. p. 61.

28. Venus. . . . . Bivalve, anterior margin incumbent. Hinge with three teeth. Vulva and anus distinct. Vol. I. p. 199. *Syst. Nat.* p. 1128. *Brand. fos. hant.* fig. 90, 91, 93, 94, 104, 105.

29. Cardium. *Cockle.* Bivalve, equivalve, with two middle teeth, alternate; lateral teeth remote, inserted. Vol. I. p. 198. *Syst. Nat.* p. 1121. *Brand fos. hant.* fig. 92, 96, 98, 99. Heart-shaped. Clay-pit at Richmond in Surry; Shereborn, Glocestershire; Harwich Cliff; Shooter's-Hill; and in huge masses of grey lime-stone near Castle Saffron in the county

county of Cork. *Smith*, vol. II. p. 374.
Woodw. tom. I. part ii. p. 53.

30. Mya. .... Bivalve, open at one end; hinge with a broad thick tooth, not let into the opposite shell. Vol. I. p. 197. *Syst. Nat.* 1112. *Brand. fof. hant.* fig. 95. *List. Angl.* tom. II. f. 30.

31. Arca. .... Bivalve, equivalve; teeth of the hinge numerous, acute, alternate, inserted. Harwich Cliff; Shotover hill, Oxfordshire; various parts of Glocestershire. *Lin. Syst. Nat.* p. 1140. *Brand. fof. hant.* fig. 97. 101. 106. *Woodw.* tom. I. part ii. p. 52.

32. Solen. .... Bivalve, oblong, open at each side; hinge with a single or double conic bent tooth, not inserted in the opposite valve. Vol. I. p. 197. *Syst. Nat.* p. 1113. *Brand. fof. hant.* fig. 103. *Woodw.* tom. I. part ii. p. 63. In many parts of Glocestershire.

33. Mytilus. *Muscle.* Bivalve, equivalve, oblong, rough; hinge without teeth, distinct, with a fubulated hollow longitudinal line. In various parts of Glocestershire. Vol. I. p. 201. *Brand. fof. hant.* fig. 124. *Syst. Nat.* p. 1155. *Luid.* cap. 8. *Smith's Cork*, vol. II. p. 380. *Woodw.* tom. I. part ii. p. 58. 62.

34. Patella. *Limpet.* Shell single, subconic, not voluted. In stone quarries in Glocestershire,

shire, Harwich Cliff, &c. but very rare, Vol. I. p. 207. *Woodw.* tom. I. part ii. p. 23. *Wall. Spec.* 370.

35. Cypræa. A single shell, involuted, oval, obtuse, smooth; aperture dentated, longitudinal. Vol. I. p. 202. *Syst. Nat.* 1172. *Woodw.* tom. I. part ii. p. 35. §. 3. In a clay-pit at Richmond, Surry.

36. Pholas. Bivalve, with one or more irregular small valves at the hinge, which is recurved. Vol. I. p. 196. Harwich Cliff. *Woodw.* tom. I. part ii. p. 63.

PHYTOLITHUS. *Vegetables.* Vol. II.

1. Plantæ. *Grass, Reeds, Horsetail,* &c. Found frequently in the black slate called plate, immediately above the pit-coal, in various parts of England. See Shistus. Also in detached nodules. *Lin.* 42, 1. *Brand. fos. hant.* fig. 122. *Woodw.* tom. I. part ii. p. 10, 11.

2. Filices. *Ferns.* Found frequently in the black slate (see Shistus) above the pit-coal, at Newcastle and in other parts of England. *Lin.* 42, 2. *Luid.* cap. 3. *Woodw.* tom. I. part ii. p. 9, 12.

3. Lithoxylon. *Petrefied wood.* Found buried in the earth, and sometimes in lakes, as in Lough Neagh in Ireland. *Lin.* 42, 4. *Luid.* cap. 4. *Woodw.* tom. I. part ii.

ii. p. 20. Harwich Cliff, *Woodw.* tom. II. p. 57.

4. Rhizolithus. *Roots of trees and plants.* Found buried in the earth. *Lin.* 42, 3. *Woodw.* tom. I. part ii. p. 18.

5. Lithophyllum. *Leaves of trees*, particularly Oak, found petrefied or incrufted in water impregnated with calcareous matter; as at Knaresborough in Yorkfhire, &c. Alfo impreffed in ftone. *Lin.* 42, 5.

6. Carpolithus. *Fruits*, particularly impreffions of the cones of pines, hazel, oak. *Lin.* 42, 7. *Woodw.* tom. I. part ii. p. 16, 21. tom. II. p. 92.

ZOOLITHUS. *Mammalia.* Vol. I. Clafs I.

1. Cervi. *Stags horns*, particularly of the moufe-deer, often found buried in the ground in fome mountains in England and Ireland. *Woodw. Meth.* 124. *Cat.* tom. I. part ii. p. 86.

2. Elephantis. *Elephants tufks*, grinders, bones, &c. *Woodw. Meth.* 124. *Cat.* part ii. p. 86.

3. Turcofa. *Bone* tinged green by copper, found in copper mines in Cumberland. *Woodw.* tom. I. part ii. p. 87.

4. Os. *Animal bones.* Brand. *fof.* fig. 118, 119, 120.

120, 121. *Woodw.* tom. I. part ii. p. 87.

AMPHYBIOLITHUS. *Amphibia.* Vol. I. Clafs III. *Lin. Syft. Nat.* Vol. III. gen. 38.

1. Gloſſopetræ. *Sharks teeth.* Somewhat reſembling a tongue with the root, or like the head of an arrow; black, or bluiſh, or brown, from half an inch to four inches in length, poliſhed. Found on the Kentiſh coaſt, alſo in ſtrata of clay at Richmond in Surry, at Harrow on the Hill, at Highgate, Iſlington, &c. *Luid.* cap. 19. *Brand. foſ. hant.* fig. 111, 112, 113, 114, 115. *Woodw.* tom. I. part ii. p. 83. Found alſo at Whitney in Oxfordſhire.

2. Plectronites. *Cock Spurs.* The teeth of an unknown fiſh. Conical, ſharp-pointed, without root, often bent; from a quarter of an inch to two inches long; or brown, or grey, or black, poliſhed. Frequently found in ſtone quarries in many parts of England. *Luid.* cap. 19. *Woodw.* tom. I. part ii. p. 84.

ICHTHYOLITHUS. *Fiſhes.* See Vol. I. Clafs VI.

1. Totalis. Impreſſion of an intire flat fiſh found in blackiſh ſlate in Wales. *Luid.* Ep. I. tab.

tab. 22. fig. 2. *Lin. Syst. Nat.* vol. III. gen. 39, 1.

2. Bufonites. *Molares of the Sea-wolf.* Vol. I. p. 68. Usually roundish and hollowed like a cup, from the size of a small pea to near an inch in diameter; black, or grey, or brown, sometimes variegated, always finely polished. *Wall. Spec.* 350. *Luid.* cap. 20. In various parts of England, particularly near Whitney in Oxfordshire. *Woodw.* tom. I. part ii. p. 84.

3. Siliquastra. *Fossile Pods.* Supposed to be the bony palates of different fishes, often resembling half the pod of the lupine, or other leguminous plant, filled with stony matter, sometimes extremely minute, and sometimes near two inches long; or brown, or black, or bluish. Frequently found in stone quarries near Shereborn in Glocestershire, Whitney in Oxfordshire, Grafton in Northamptonshire, Farrington in Berkshire, &c. *Brand. fos. hant.* fig. 116, 117. *Luid.* n. 1440. *Woodw.* tom. I. part ii. p. 85.

4. Vertebra. Vertebræ of fishes of various genera, often found in pits and quarries in different parts of the kingdom, particularly at Richmond in Surry; also on the cliffs of Sheppy Island, Pyrton Passage, &c. *Luid.* cap. 22. *Brand. fos. hant.*

fig. 108, 109, 110. *Woodw.* tom. I. part ii. p. 82.

ENTOMOLITHUS. *Insects.* Vol. I. Class V. p. 85.

1. Cancri. *Crabs.* Claws or parts of claws, found in pits, &c. in several parts of England. *Luid.* cap. 18. Two small crabs found in the cliffs at Folkstone near Dover. *Woodw.* tom. I. part ii. p. 81.

GRAPTOLITHUS. Stones on which various figures are depicted by some fossile fluid or vapour.

1. Dendrites. Representing shrubs, or plants, or moss. On various stones, slates, and flints, found in different parts of England, particularly on a whitish stone in Sella Park in Cumberland. *Woodw.* m. I. p. 239.

CLASS. VI.

WATER.

Not inflammable; very little, if at all, compressible; soluble in air; by heat rendered volatile and extremely elastic; fluid when Fahrenheit's thermometer stands above 32 degrees, and solid when it is below that point.

> PURE or COMMON WATER. Pellucid, colourless, inodorous, insipid. Dissolves salts, gums, mucilages. Miscible with vinous spirits. Specific gravity to gold as 1000 to 19,640.

1. *Dew.* Supposed to be the lightest and most pure.

2. *Rain.* Next to dew in purity; it is nevertheless impregnated with variety of heterogeneous matter, particularly calcareous earth and some neutral salts. *Berlin Mem.* An. 1751. p. 131.

3. *Snow, Hail, Ice.* Equally pure, if not more so, than rain-water; its contents nearly the same.

4. *Spring-water.* Differently impregnated, according to the soil through which they pass,
with

with selenites, earth, and other fossile matter.

5. *River-water.* Besides fossile impregnations, contains great variety of animal and vegetable substances.

6. *Pond-water.* Contains most animal and vegetable matter; therefore most liable to putrefaction, and consequently least fit for use.

ALKALINE WATER. Impregnated with fossile alkali *per se*. Effervesces with all acids; changes syrup of violets and other vegetable blues to green; precipitates solutions of calcareous earths in acids; and of sal ammoniac or alum in water; becomes opaque and white with solutions of silver, lead, or mercury, in the nitrous acid; precipitates iron from the vitriolic or nitrous acid.

1. Clifton, *Oxfordshire.* Limpid, with very little taste. A gallon yields about 70 grains of residuum. Contains, besides fossile alkali, a small proportion of calcareous earth, and another salt, probably *sal cath. amar. Rutty. Syn.* p. 429. *Short*, vol. II. p. 133. *Monro*, vol. I. p. 82.

2. Glassonbury, *Somersetshire.* Contains a small proportion of alkali and some sea-salt. A gallon yields about 30 grains of residuum

duum by evaporation. *Rutty. Monro*, vol. I. p. 84.

3. Tilbury, *Essex*. Not perfectly limpid at the well. Curdles with soap, but not with milk. Turns milky when boiled, but rendered clear by acids. Contains much air, fossile alkali, and calcareous earth. A gallon yields on evaporation about 200 grains of residuum. *Rutty. André. Monro*, vol. I. p. 78.

Ireland.

4. St. Bartholomew's Well, *Cork*. Mixes uniformly with soap. A gallon yields by evaporation about 24 grains of residuum, consisting chiefly of fossile alkali. *Monro*, vol. I. p. 85.

5. Cape Clear, *Cork*. Lathers with soap. A gallon yields about 30 grains of residuum, consisting chiefly of fossile alkali, with a small quantity of sea-salt. *Monro*, vol. I. p. 86.

6. Carrick-moor, *Cavan*. Tastes soft, like Bristol water; curdles soap, and deposites a white sediment with *lix. tartari*. Contains an alkaline salt, *sal cath. amar.* and calcareous earth, probably deprived of its fixt air. *Rutty. Monro*, vol. I. p. 85.

7. Tober Bony, *Dublin*, four miles north. Lathers

thers easily with soap. A gallon yields
about 20 grains of residuum, consisting
of an alkaline salt and calcareous earth.
Rutty. Monro, vol. I. p. 84.

SALT WATER. Impregnated with a
predominant sea-salt. Precipitates solutions of silver, lead, or mercury, in a
white cloud. Its salts easily distinguished, on evaporation, by their cubical crystals. When simply impregnated, is in
no respect affected by acids, alkalis,
vegetable astringents, or syrup of violets.
Decomposed by the vitriolic or nitrous
acid.

1. Sea-Water. Contains sea-salt in large proportion, calcareous earth, bittern, and a
small quantity of oil.

England.

2. Barrow-dale, *Cumberland.* Contains a large
proportion of sea-salt, some calcareous
earth, and a little Epsom salt. *Short*,
vol. II. p. 85. *Rutty. Monro*, vol. I.
p. 118.

3. Dortshill, *Staffordshire.* Contains sea-salt, calcareous earth, and bittern. *Short*, 8vo.
1765. *Monro*, vol. I. p. 122.

4. St. Erasmus's Well, *Staffordshire*, on Lord
Chetwynd's estate. Colour of sack, with
little taste or smell. Contains a small
quantity

quantity of sea-salt, some earth, and a little Epsom salt. *Short. Rutty. Monro*, vol. I. p. 121.

5. Leamington, *Warwickshire*. Contains a small proportion of sea-salt, a little calcareous earth, and also some *sal cath. amar. Short*, vol. II. p. 87. 133. *Rutty. Monro*, vol. I. p. 119.

6. Rougham, *Lancashire*. Contains a small quantity of sea-salt, some calcareous earth, *sal cath. amar.* and natron, *Short*, vol. II. p. 85. 132, *Rutty. Monro*, vol. I. p. 120.

7. Salt Springs, *Cheshire*. Northwich, Droitwich, Upwich, Middlewich, Namptwich; Barton in Lancashire, Weston in Staffordshire, &c. *Short*, vol. II. p. 85. *Monro*, vol. I. p. 112. Contain a very large proportion of sea-salt.

Wales.

8. Cargyrle, *Flintshire*. Contains a small proportion of sea-salt, a little calcareous earth, and *sal cath. amar. Short*, vol. II. p. 86. *Rutty. Monro*, vol. I. p. 121.

Ireland.

9. Carrickfergus, *Antrim*. Contains some sea-salt, insoluble earth, and a little Epsom salt. *Monro*, vol. I. p. 122.

10. Kilroot,

10. Kilroot, *Antrim.* Contains sea-salt, a large proportion of insoluble matter, and some little Epsom salt. *Monro*, vol. I. p. 123.

11. Mahereberg, *Kerry.* Contains sea-salt and Epsom salt. *Rutty. Smith's Nat. Hist. of Kerry. Monro*, vol. I. p. 123.

CATHARTIC WATER. Impregnated with a predominant bitter purging salt, commonly called by writers on mineral waters, Calcareous Glauber's salt, or, with more propriety, Magnesia Glauber's salt, or Epsom salt; composed of vitriolic acid and *magnesia alba*. Precipitates solutions of silver, lead, or quicksilver, in the nitrous acid, in a yellow cloud. Is not affected by acids, but precipitates with an alkali. Crystals resemble those of Glauber's salt.

England.

1. Alford, *Somersetshire.* A gallon contains about four scruples of Epsom salt, about half that quantity of sea-salt, and one scruple of calcareous earth. *Guidot. Rutty. Monro*, vol. I. p. 132.

2. Acton, *Middlesex.* Tastes bitter and saltish; a powerful cathartic. Contains Epsom salt, and probably selenites, besides some calcareous earth. *Rutty. Monro*, vol. I. p. 144.

3. Alkerton,

3. Alkerton, *Glocestershire.* Contains a large proportion of Epsom salt, calcareous earth, and a little sea-salt. Short, 8vo, 1765. *Monro,* vol. I. p. 148.

4. Bagnigge, *Middlesex.* A brisk purgative. Contains Epsom salt, sea-salt, and calcareous earth. *Bevis Experim. Enquiry,* 1760. *Monro,* vol. I. p. 142.

5. Barnet, *Hertfordshire.* Contains a large proportion of Epsom salt, a little sea-salt, and some insoluble earth. *Rutty. Monro,* vol. I. p. 143.

6. Ball-well, *Lincolnshire,* at Stenfield. Contains Epsom salt in small quantity, a considerable proportion of earth, and some sea-salt. *Short,* vol. I. p. 107. *Monro,* vol. I. p. 149.

7. Comner, *Berkshire.* Colour whitish. Contains Epsom salt, some calcareous earth, and, according to Rutty, a mixture of natron. *Short,* vol. II. p. 80. *Monro,* vol. I. p. 141.

8. Dog and Duck, *Surry.* A weak cathartic. Contains Epsom salt and sea-salt, with one twelfth of the residuum of insoluble matter. *Hale's Phil. Transf.* N°. 493. *Rutty,* p. 168. *Monro,* vol. I. p. 136.

9. Dulwich, *Kent.* Clear, somewhat brackish, and

and bitter. Contains Epsom salt and sea-salt in nearly equal proportion, and a little calcareous earth. *Rutty*, p. 170. *Monro*, vol. I. p. 133.

10. Epsom, *Surry*. Limpid, with a slight saline taste. Contains *sal cath. amar.* in large proportion, calcareous earth, and probably selenites. *Rutty. Lucas. Allen. Monro*, vol. I. p. 146.

11. Hanlys, *Shropshire*. A strong cathartic. Tastes salt and bitterish; contains a large proportion of Epsom salt, and some insoluble earth. *Linden Hist.* 1768. *Monro*, vol. I. p. 140.

12. Holt, *Wiltshire*. Limpid, with very little taste. Contains Epsom salt and calcareous earth in nearly equal proportion, and a little sea-salt. *Rutty. Monro*, vol. I. p. 134.

13. Kinalton, *Nottinghamshire*. Clear, saltish. Contains Epsom salt, and a very pure calcareous earth. *Rutty. Monro*, vol. I. p. 138.

14. Morton-see, *Shropshire*. A mild purgative. Contains Epsom salt and some calcareous earth. *Short*, vol. II. p. 81.

15. North-hall, *Hertfordshire*, near Barnet. Contains a large proportion of Epsom-salt, a little sea-salt, and some calcareous earth. *Rutty. Monro*, vol. I. p. 143.

16. Pancras,

16. Pancras, *Middlesex*. A mild purgative. Contains Epsom salt, a little sea-salt, and some insoluble earth. *Monro*, vol. I. p. 142.

17. Stretham, *Surry*. Curdles with soap, and also with milk when boiled with it. Contains Epsom salt, sea-salt, and selenites. *Rutty. Monro*, vol. I. p. 135.

18. Sydenham, *Kent*. A mild cathartic. Contains Epsom salt, sea-salt, and some calcareous earth. *Rutty. Monro*, vol. I. p. 138.

19. Nevil Holt, *Leicestershire*, near Market-Harborough. Contains a confiderable proportion of Epsom salt, some calcareous earth, selenites, fixt air, vitriolic acid, iron, and possibly a little alum. *Short*, octavo, 1765. p. 156. *Monro*, vol. I. p. 436. *Rutty*.

Wales.

20. Llandrindod, *Radnor*. Contains Epsom salt, sea-salt, and some earth. *Linden. Monro*, vol. I. p. 149.

Ireland.

21. Carrickfergus, *Antrim*. Colour bluish; contains a small quantity of Epsom salt, some calcareous earth, and a little sea-salt. *Rutty. Monro*, vol. I. p. 151.

SULPHUR WATER. Water containing Sulphur *per se* diffused, or *hepar sulphuris* dissolved; of which last kind are most, if not all, the sulphur-waters in these kingdoms. Strikes a black colour with a solution of lead in the nitrous acid, or of *sacc. saturni* in water; tarnishes silver; smells like the washing of a foul gun; becomes milky with acids.

England.

1. Askeron, *Yorkshire*, near Doncaster. Perfectly clear; contains much sulphur, Epsom salt, a little sea-salt, and a large proportion of earth. *Short*, vol. I. p. 303. *Monro*, vol. I. p. 212.

2. Bilton, *Yorkshire*, near Knaresborough. Contains sulphur, natron, a little sea-salt, and some earth. *Short*, vol. I. p. 296. *Rutty*. *Monro*, vol. I. p. 181.

3. Broughton, *Lancashire*. Contains a considerable proportion of sulphur, sea-salt, Epsom salt, and earth. *Short*, vol. I. p. 300. *Rutty*. *Monro*, vol. I. p. 200.

4. Buglawton, *Cheshire*, near Congleton. Extremely cold: Contains sulphur, a small proportion of Epsom salt, and a little calcareous earth. *Short*, vol. II. p. 62. *Monro*, vol. I. p. 215.

5. Chadlington, *Oxfordshire.* Tastes saltish; contains sulphur, natron, a little sea-salt, and some earth. *Short,* vol. II. p. 70. *Rutty. Monro,* vol. I. p. 181.

6. Crickle, *Lancashire,* near Braughton. Contains much sulphur, sea-salt, Epsom salt, and some calcareous earth. *Short,* vol. I. p. 300. *Rutty. Monro,* vol. I. p. 199.

7. Cunley-house, *Lancashire,* near Whaley. Colour bluish. Contains sulphur, magnesia Glauber's salt, and earth. *Short,* vol. II. p. 60. *Monro,* vol. I. p. 214.

8. Codsall Wood, *Staffordshire,* near Wolverhampton. Contains a considerable quantity of sulphur, with very little calcareous earth. *Short,* vol. II. p. 63. *Monro,* vol. I. p. 218.

9. Croft, *Yorkshire,* near the bishopric of Durham. Clear, sparkling. Contains sulphur, much calcareous earth, some Epsom salt, and a little sea-salt. *Short,* vol. I. p. 299. vol. II. p. 134.

10. Cawley, *Derbyshire,* near Dranefield. Contains sulphur, Epsom salt, and a little calcareous earth. *Short,* vol. I. p. 305. *Monro,* vol. I. p. 213.

11. Durham, on the north side of the river. Contains sulphur, some sea-salt, and a little

little earth. *Short*, vol. I. p. 305. *Monro*, vol. I. p. 202.

12. Deddington, *Oxfordshire*, near Banbury. Contains sulphur, iron, and sea-salt, according to *Short*; but in *Rutty*'s opinion it is a fossile alkali. If that were the case, how happens this fossile alkali not to expel the iron from its acid, and form Glauber's salt?

13. Drig-well, *Cumberland*, near Ravenglas. A clear, brisk water, containing sulphur and iron. *Short*, vol. II. p. 63. *Monro*, vol. I. p. 453.

14. *Gainsborough*, Lincolnshire. Contains sulphur, a little iron, and some magnesia Glauber's salt. *Short*, vol. II. p. 69. *Monro*, vol. I. p. 454.

15. Harrigate, *Yorkshire*. Contains a considerable proportion of sulphur and sea-salt, a little magnesia Glauber's salt, and some earth. *Short*, vol. I. p. 285. *Monro*, vol. I. p. 193.

16. Keddleston, *Derbyshire*. Contains a large proportion of sulphur, with sea-salt and calcareous earth. *Short*, vol. I. p. 305. *Monro*, vol. I. p. 201.

17. Loansbury, *Yorkshire*, in Lord Burlington's park. Contains a little sulphur, Epsom salt,

falt, and earth. *Short*, vol. II. p. 61. *Monro*, vol. I. p. 215.

18. Maudsley, *Lancashire*, near Preston. Colour bluish; faltish taste; contains a large proportion of sulphur, sea-salt, and a little calcareous earth. *Short*, vol II. p. 63. *Rutty*. *Monro*, vol. I. p. 198.

19. Nottington, *Dorsetshire*, near Weymouth. Contains sulphur, natron, and a little earth. *Rutty*, p. 519. *Monro*, vol. I. p. 183.

20. Normanby, *Yorkshire*, near Pickering. Clear, but covered with a blue scum. Contains much mephitic air, some sulphur, a little bitter purging salt, and some sea-salt. *Short*, vol. I. p. 299. *Monro*, vol. I. p. 210.

21. Quin Camel, *Somersetshire*. Contains sulphur, natron, a little sea-salt, and some earth. *Rutty*. *Monro*, vol. I. p. 182.

22. Rippon, *Yorkshire*. Contains sulphur, sea-salt, Epsom salt, and a large proportion of earth. *Short*, 8vo, 1765, p. 72. *Monro*, vol. I. p. 203.

23. Sutton-bog, *Oxfordshire*. A mild cathartic; smells extremely fœtid; tastes saltish; throws up a blue scum. Contains sulphur, fossile alkali, some sea-salt, and a little

little earth. *Short*, vol. II. p. 70. *Rutty, Monro*, vol. I. p. 179.

24. Shattlewood, *Derbyshire*. Contains a little sulphur, a good deal of sea-salt, and a small quantity of earth. *Short*, vol. I. p. 304.

25. Skipton, *Yorkshire*. Contains sulphur, sea-salt, Epsom salt, with some earth. *Short*, 8vo, 1765, p. 71. *Monro*, vol. I. p. 203.

26. Shapmoor, *Westmoreland*, between Shap and Orton. Contains much sulphur and Epsom salt, with a little sea-salt and earth. *Rutty. Monro*, vol. I. p. 217.

27. Thorparch, *Yorkshire*, above Tadcaster, on the river Wharf. Contains a little sulphur, iron, sea-salt, and earth. *Monro*, vol. I. p. 454.

28. Upminster, *Essex*, near Brentwood. Contains much sulphur, Epsom salt, and natron, (according to Dr. Rutty) with some earth. *Rutty. Monro*, vol. I. p. 218.

29. Wigglesworth, *Yorkshire*, near Settle. Black, covered with a white scum. Tastes saltish; lathers with soap. Contains sulphur, natron, a little sea-salt, and some black earth. *Rutty. Short*, vol. I. p. 302. *Monro*, vol. I. p. 180.

30. Wardrew,

30. Wardrew, *Northumberland*, on the river Arden. Contains a very confiderable proportion of fulphur, a fmall quantity of fea-falt, and very little earth. *Short*, vol. II. p. 62. *Monro*, vol. I. p. 202.

31. Wirkfworth, *Derbyſhire*. Colour black. Contains a little fulphur, bitter falt, and iron. *Short*, vol. I. p. 307. *Monro*, vol. I. p. 219.

Ireland.

32. Anaduff, *Leitrim*. Contains fulphur, natron, calcareous Glauber's falt, (as we are told) with fome calcareous earth. *Monro*, vol. I. p. 189.

33. Aphaloo, *Tyrone*. Contains fulphur, natron, and (according to Rutty) calcareous, or rather magnefia Glauber's falt, with a fufpicion of a fmall chalybeate impregnation. *Monro*, vol. I. p. 189.

34. Aſhwood, *Fermanagh*. Contains fulphur, natron, and Epfom falt, as we are told. *Monro*, vol. I. p. 187.

35. Ballynahinch, *Down*. A very clear chalybeate, containing fome fulphur, and a neutral falt of fome fort or other. *Rutty*. *Monro*, vol. I. p. 455.

36. Caſtle-

36. Caftlemaign, *Kerry*. Contains fulphur, iron, and a falt of fome fort. *Rutty*. *Monro*, vol. I. p. 455.

37. Drumgoon, *Fermanagh*. Contains fulphur, natron, a little fea falt, and fome infoluble matter. *Rutty*. *Monro*, vol. I. p. 183.

38. Derrylefter, *Cavan*. Somewhat lighter than common water. Contains a large proportion of fulphur, fome natron, very little fea falt, and fome earth. *Rutty*. *Monro*, vol. I. p. 185.

39. Derryhence, *Fermanagh*. Contains fulphur, natron, and fea-falt. *Rutty*. *Monro*, vol. I. p. 187.

40. Drumafnave, *Leitrim*. Contains a confiderable proportion of fulphur, fome natron, Epfom falt, and infoluble earth. *Rutty*. *Monro*, vol. I. p. 188.

1. Derrindaff, *Cavan*. Contains fulphur, fome Epfom falt, and a little earth. *Rutty*. *Monro*, vol. I. p. 220.

42. Killafher, *Fermanagh*. Contains fulphur in confiderable proportion, fome natron, and alfo fea-falt, together with *fal cath. amarum*, according to Dr. *Rutty*. *Monro*, vol. I. p. 186.

43. Lifbeak,

43. Lisbeak, *Fermanagh*. Contains sulphur, natron, and a little earth. *Rutty*. *Monro*, vol. I. p. 186.

44. Mechan, *Fermanagh*. Contains sulphur, natron, and sea-salt. *Rutty*. *Monro*, vol. I. p. 187.

45. Owen Breun, *Cavan*. Contains sulphur, Epsom salt, and a small portion of natron, (according to Dr. *Rutty*) with some calcareous earth, and something else. *Monro*, vol. I. p. 220.

46. Pettigoe, *Donegal*. Contains a large proportion of sulphur, Epsom salt, and a little earth. *Rutty*. *Monro*, vol. I. p. 221.

47. Swadlingbar, *Cavan*. Generally covered with a bluish scum. Contains sulphur, natron, Epsom salt, and calcareous earth. *Rutty*. *Monro*, vol. I. p. 184.

Wales.

48. Llandrindod, *Radnor*. Contains sulphur and sea-salt. *Linden*, *Monro*, vol. I. p. 204.

Scotland.

49. Carstarphin, near Edinburgh. Contains a little sulphur, sea-salt, and Epsom salt, with

with some earth. *Short. Monro*, vol. I. p. 209.

50. Moffat, *Annandale.* Bluish colour; contains a considerable proportion of sulphur, sea-salt, and a little earth. *Plummer. Med. Essays*, vol. I. art. 8. *Monro*, vol. I. p. 105.

COPPER WATER: Impregnated with copper dissolved in vitriolic acid. Turns blue with caustic volatile alkali. Precipitated by iron.

Ireland.

1. Ballymurtogh, *Wicklow.* A gallon yields seven drachms and a half of sediment on evaporation, from which were obtained green and bluish crystals, and also a little white vitriol, according to Dr. *Rutty. Monro*, vol. I. p. 235.

2. Cronebaun, *Wicklow.* Near the last, but on the opposite side of the river Arklow. Yielded, on evaporation of a gallon, four drachms and sixteen grains of sediment, which contained blue and green vitriol. *Rutty. Monro*, vol. I. p. 236.

CHALYBEATE WATER. Iron dissolved in the vitriolic acid. Strikes a black

black or purple colour with vegetable aftringents, especially with the addition of an alkali or lime water. Precipitated by quickfilver, diffolved in the nitrous acid; alfo by alkaline falts.

England.

1. Aftrope, *Oxfordshire*, near Banbury. Clear, brifk; contains iron with a little Epfom falt, and fome earth. *Short*, vol. II. p. 45. *Monro*, vol. I. p. 376.

2. Afwerby, *Lincolnshire*. Colour bluifh; contains iron, a large proportion of Epfom falt, and fome calcareous earth. *Short*, vol. I. p. 217. *Monro*, vol. I. p. 388.

3. Birmingham, *Warwickshire*. A brifk chalybeate, with little folid contents. *Short*, vol. II. p. 43. *Monro*, vol. I. p. 274.

4. Buxton, *Derbyshire*. A cold chalybeate water, containing alfo a little fea-falt and *fal cath. amarum*. *Short*, vol. I. p. 229. *Monro*, vol. I. p. 357.

5. Burlington, *Yorkshire*. Contains iron, a little Epfom falt, and calcareous earth. *Short*, vol. I. p. 230. *Monro*, vol. I. p. 375.

6. Bournley,

6. Bournley, *Lancashire.* A light chalybeate, with a little Epsom salt, and probably some selenites. *Short,* vol. II. p. 130. *Monro,* vol. I. p. 277.

7. Binley, *Warwickshire,* near Coventry. A very light chalybeate, containing some Epsom salt. *Short,* vol. II. p. 45. *Monro,* vol. I. p. 378.

8. Bagnigge, *Middlesex,* near London. Contains iron, Epsom salt, earth, and some selenites. *Bevis. Monro,* vol. I. p. 399.

9. Carlton, *Nottinghamshire.* A very light water impregnated with iron dissolved in vitriolic acid. *Short,* vol. II. p. 40. *Monro,* vol. I. p. 266.

10. Colurian, *Cornwall.* A chalybeate water the contents of which are not well known. *Borlase. Monro,* vol. I. p. 271.

11. Cannock, *Staffordshire.* A very light chalybeate. *Short,* vol. II. p. 43. *Monro,* vol. I. p. 274.

12. Cobham, *Surry.* Contains iron with a small quantity of sea-salt. A gallon yields but seven grains of residuum. *Monro,* vol. I. p. 355.

13. Chippenham, *Wiltshire.* Contains iron and sea-salt. *Rutty. Monro,* vol. I. p. 359.

14. Cawthorp, *Lincolnshire,* near Bounre. Contains iron, with a large proportion of sea-salt, a little Epsom salt, and probably some selenites. *Short.* vol. I. p. 225. *Monro,* vol. I. p. 362.

15. Coventry, *Warwickshire.* A light chalybeate, containing also a little Epsom salt. *Short. Monro,* vol. I. p. 376.

16. Cheltenham, *Glocestershire.* Contains iron dissolved in a volatile acid, a large proportion of Epsom salt and calcareous earth, some selenites, and probably fixt air. *Lucas. Rutty. Short. Monro,* vol. I. p. 395.

17. Derby, *Derbyshire.* A strong chalybeate, containing also a large proportion of sea-salt. *Short,* vol. II. p. 48. *Monro,* vol. I. p. 362.

18. Dorsthill, *Staffordshire.* A brisk chalybeate, containing also sea-salt, Epsom salt, and bittern. *Monro,* vol. I. p. 365.

19. Felstead, *Essex.* A light chalybeate. *Allen. Monro,* vol. I. p. 269.

20. Filah,

20. Filah, *Yorkshire*, near Scarborough. Colour whitish; contains, besides iron, a considerable proportion of sea-salt, some Epsom salt, calcareous earth, and probably much fixt air. *Short*, vol. I. p. 289. *Monro*, vol. I. p. 364.

21. Hampstead, *Middlesex*. A pure chalybeate, containing a small proportion of iron dissolved in a fixt vitriolic acid, with probably a little sea-salt. *Soam*, 1734. *Boyle*. *Monro*, vol. I. p. 264.

22. Harrigate, *Yorkshire*. *Sweet spaw:* Lighter than common water; contains iron dissolved in vitriolic acid partly fixt, with some earth. *Tuewhet well:* Lighter than the former, containing more iron, dissolved in a more volatile acid, together with a greater proportion of earth. *Short*. *Monro*, vol. I. p. 271.

23. Hartlepool, *Bpk. of Durham*. Contains iron in a volatile acid, with a good deal of Epsom salt and calcareous earth; also some sea-salt, and possibly a little sulphur. *Short*, vol. II. p. 59. *Monro*, vol. I. p. 380.

24. Hanlys, *Shropshire*, near Shrewsbury. Contains iron, with a large proportion of Epsom salt and calcareous earth. *Linden*. *Monro*, vol. I. p. 402.

25. Islington,

25. Iſlington, *Middleſex.* A very light water impregnated with iron diſſolved in a volatile vitriolic acid. A gallon yields about a ſcruple of reddiſh earth. *Boyle. Linden. Monro*, vol. I. p. 267.

26. Ilmington, *Warwickſhire.* Clear, briſk. Contains iron and a ſalt, concerning which authors differ. *Durham*, 1685. *Short*, vol. II. p. 129. *Monro*, vol. I. p. 302.

27. Jeſſop, *Surry.* A weak chalybeate, containing a very large proportion of Epſom ſalt, and poſſibly a little ſea-ſalt. *Hales, Phil. Tranſ.* Numb. 495. *Rutty. Monro*, vol. I. p. 400.

28. Knowſley, *Lancaſhire.* Contains iron, with a very little Epſom ſalt, and probably ſome fixt air, with a ſmall proportion of ſelenites. *Short*, vol. II. p. 129. *Monro*, vol. I. p. 375.

29. King's Cliff, *Northamptonſhire.* Contains iron, with much Epſom ſalt, and ſome earth. *Short. Rutty. Monro*, vol. I. p. 379.

30. Kerby, *Weſtmoreland*, near Appleby. Contains iron, Epſom ſalt, with a good deal of earth. *Short*, vol. II. p. 132. *Rutty. Monro*, vol. I. p. 384.

31. Leez,

31. Leez, *Essex*. Impregnated with iron dissolved in a volatile vitriolic acid. *Allen. Monro*, vol. I. p. 268.

32. Lincomb, *Somersetshire*, near Bath. Contains iron, and, according to Dr. Hillary, both natron and Epsom salt. *Monro*, vol. I. p. 305.

33. Latham, *Lancashire*. Clear, containing iron, sea-salt, and earth. *Short. Monro*, vol. I. p. 358.

34. Lancaster, *Lancashire*. Contains iron, sea-salt, and probably selenites. *Short*, vol. II. p. 130. *Monro*, vol. I. p. 360.

35. Marks-hall, *Essex*. Becomes red with galls, which colour is said to disappear in two days, but without any precipitation. *Allen. Monro*, vol. I. p. 268.

36. Malvern, *Glocestershire*. Contains iron dissolved in a volatile vitriolic acid. Evaporated two quarts yielded one grain of earth, one of iron, and one of bittern. *Wall. Monro*, vol. I. p. 270.

37. Moss House, *Lancashire*, near Maudsley. A brisk light chalybeate; a gallon yielded on evaporation 23 grains of residuum, five of which were a salt of some sort or other.

other. *Short,* vol. II. p. 38. *Monro,* vol. I. p. 274.

38. Malton, *Yorkshire.* A strong chalybeate, containing also a considerable quantity of Epsom salt and earth, with fixt air and selenites. *Lister. Short. Monro,* vol. I. p. 386.

39. Newham Regis, *Warwickshire.* Contains iron and Epsom salt, with some calcareous earth. *Short. Monro,* vol. I. p. 378.

40. Orston, *Nottinghamshire.* Clear, pleasant, containing iron dissolved probably by means of fixt air, with a large proportion of earth, together with Epsom salt and some sea-salt. *Short,* vol. I. p. 222. *Monro,* vol. I. p. 382.

41. Road, *Wiltshire.* Contains iron, and, according to the following Doctors, a large quantity of natron, also some sea-salt. *Williams. Clark. Rutty. Monro,* vol. I. p. 306.

42. Shadwell, *Middlesex,* near London. A very strong chalybeate; a gallon yields two ounces and three drachms of salt of steel, and three drachms of a yellowish brown earth. *Rutty. Monro,* vol. I. p. 247.

43. Sene,

43. Sene, or Send, *Wiltshire*, near the Devizes. A strong chalybeate: Its acid seems fixt. *Guidott, an.* 1691. p. 405. *Monro*, vol. I. p. 275.

44. Stanger, *Cumberland*, two miles from Cockermouth. Contains a good deal of iron, a large quantity of sea-salt, and probably some selenites. *Short*, vol. II. p. 132. *Monro*, vol. I. p. 366.

45. Stenfield, *Lincolnshire*. Clear, light, brisk. Contains iron in a fixt acid, with a large proportion of Epsom salt and earth, together with a little sea-salt, and probably some selenites. *Short*, vol. I. p. 214. *Monro*, vol. I. p. 383.

46. Scarborough, *Yorkshire*. Contains iron in a volatile acid, with a considerable proportion of Epsom salt, and calcareous earth, some selenites and fixt air. *Shaw, Enq. an.* 1734. *Atkins*. *Short*, vol. I. p. 174. *Lucas*. *Monro*, vol. I. p. 389.

47. Stockport, *Lancashire*. A gallon yielded, on evaporation, twelve grains of ochre, and the same quantity of a mixture of sea-salt and Epsom salt. *Short*, vol. II. p. 130. *Monro*, vol. I. p. 357.

48. Tunbridge, *Kent*. Contains iron dissolved in a volatile vitriolic acid, some sea-salt, with

with a little selenites and calcareous earth. According to Dr. Lucas, a gallon yields 30 grains of residuum. *Lucas. Lister. Rutty. Monro*, vol. I. p. 355.

49. Thetford, *Norfolk*. Contains iron, and, according to the following Doctor, a pure alkaline salt. *Manning, de aq. min. Monro*, vol. I. p. 304.

50. Tibshelf, *Lancashire*. Contains iron in a volatile acid, with a considerable proportion of sea-salt, a little Epsom salt, and some calcareous earth. *Short*, vol. I. p. 226. *Monro*, vol. I. p. 358.

51. Townly, or Hanbridge, *Lancashire*. Contains iron and Epsom salt. *Short*, vol. II. p. 133. *Monro*, vol. I. p. 377.

52. Thursk, *Yorkshire*. Contains iron, Epsom salt, calcareous earth, a little sea-salt, and probably fixt air. *Short*, vol. I. p. 226. *Monro*, vol. I. p. 380.

53. Thornton, *Nottinghamshire*, near Newark. Contains iron, with Epsom salt and a great deal of earth. *Short. Monro*, vol. I. p. 381.

54. Tarleton, *Lancashire*. Contains a little iron, a large proportion of Epsom salt and calcareous

calcareous earth, a good deal of sea-salt, a little sulphur, and possibly some selenites. *Short*, vol. II. p. 54. *Monro*. vol. I. p. 385.

55. Westwood, *Derbyshire*, near Tandersley. Yields pure crystals of green vitriol. *Short*, vol. I. p. 283. *Monro*, vol. I. p. 248.

56. Wellenborough, *Northamptonshire*. Lighter than common water. Contains iron, &c. *Allen. Monro*, vol. I. p. 269.

57. Wigan, *Lancashire*. Contains some iron and a little Epsom salt. *Short*, vol. II. p. 30. *Monro*, vol. I. p. 275.

58. Witham, *Essex*. Contains iron, sea-salt, Epsom salt, and calcareous earth. *Taverner. Monro*, vol. I. p. 359.

59. White-acre, *Lancashire*, near Trales. Contains iron, sea-salt, and probably selenites. *Short. Monro*, vol. I. p. 360.

60. West-Ashton, *Wiltshire*. A weak chalybeate, containing a large proportion of sea-salt and Epsom salt, with some earth. *Rutty. Monro*, vol. I. p. 361.

61. Weatherslack,

61. Weatherflack, *Westmoreland*. A weak chalybeate, containing a very large proportion of sea-salt, probably a little Epsom salt, and a very small quantity of sulphur, together with some earth. *Short*, vol. II. p. 182. *Rutty. Monro*, vol. I. p. 363.

Ireland.

62. Ardarick, *Cork*, two miles and a half southeast of the city. *Smith*, vol. II. p. 276.

63. Athlone, *Roscommon*. Contains a little iron and Epsom salt. *Rutty. Monro*, vol. I. p. 404.

64. Bandon, *Cork*. A chalybeate of considerable strength. *Smith*, vol. II. p. 271.

65. Ballyvourney, *Cork*, on the north bank of the river Sullane. A strong chalybeate. *Smith*, vol. II. p. 272.

66. Bearforest, *Cork*, a mile south of Mallow. A pretty strong chalybeate. *Smith*, vol. II. p. 274.

67. Ballycastle, *Antrim*. Contains iron in a fixt acid, calcareous earth, and a little sulphur. *Monro*, vol. I. p. 279.

68. Ballyspellan, *Tipperary*, eight miles from Kilkenny. Contains a little iron in a volatile acid. *Monro*, vol. I. p. 282.

69. Castle Townshend, *Cork*, on the road to Skibbereen. A strong chalybeate, containing also a little sulphur. *Smith*, vol. II. p. 268.

70. Cronacree, *Cork*, near Doneraile. A weak chalybeate. *Smith*, vol. II. p. 272.

71. Carrignacurra. *Cork*, near Inchiguelagh. A strong chalybeate. *Smith*, vol. II. p. 272.

72. Crofstown, *Waterford*. A gallon yields, on evaporation, 40 grains of a greenish white sediment of an acrid ferruginous taste. *Rutty. Monro*, vol. I. p. 253.

73. Castlemore, *Waterford*. A gallon yields 48 grains of vitriol. *Monro*, vol. I. p. 254.

74. Coolauran, *Fermanagh*. Contains iron in a fixt acid, and a little Epsom salt. *Rutty. Monro*, vol. I. p. 258.

75. Castleconnel, *Limeric*. Contains iron, sea-salt, calcareous earth, and probably a little selenites. *Rutty. Martin. Monro*, vol. I. p. 367.

76. Drum-

76. Drumraftle, *Cork*, near Dunmanway. Contains a small proportion of iron diffolved in a volatile acid. *Smith*, vol. II. p. 268.

77. Dunnard, eighteen miles from *Dublin*. Contains iron and a very small proportion of faline matter. *Rutty*. *Monro*, vol. I. p. 281.

78. Five-mile Bridge, *Cork*, in the road to Kinfale. A strong chalybeate, containing alfo a little fulphur. *Smith*, vol. II. p. 269.

79. Glanagarin, *Cork*, near Caftlemartyr. Contains a little iron in a volatile acid. *Smith*, vol. II. p. 268.

80. Garret's-town, *Cork*. Contains iron in a fixt acid, and fome falt. *Smith*, vol. II. p. 270.

81. Glanmile, near Naul. Contains iron diffolved in a volatile acid, and a little Epfom falt. *Monro*, vol. I. p. 279.

82. Granfhaw, *Downe*. Contains much iron, fome fea-falt, and earth. *Rutty*. *Monro*, vol. I. p. 309.

83. Galway, near the town. Contains iron in a volatile acid, a large proportion of fea-falt,

salt, some Epsom salt, and selenites. *Rutty. Monro*, vol. I. p. 370.

84. Kilindonnel, *Cork*, two miles N. by E. from the city. A strong chalybeate. *Smith*, vol. II. p. 272.

85. Kilpadder, *Cork*. A pretty strong chalybeate. *Smith*, vol. II. p. 274.

86. Killbrew, *Meath*. A gallon yielded, on evaporation, 1500 grains of sediment, chiefly vitriol of iron; supposed also to contain some copper and a little alum. *Rutty. Monro*, vol. I. p. 255.

87. Kanturk, *Cork*. Contains iron, some sulphur, and probably a little Epsom salt. *Rutty. Monro*, vol. I. p. 280. *Smith*, vol. II. p. 269.

88. Kilinshanvally, *Fermanagh*. Contains iron in a fixt acid, with a little Epsom salt and calcareous earth. *Rutty. Monro*, vol. I. p. 405.

89. Lis-done-varna, *Clare*. Contains a considerable proportion of iron, and, as some imagine, a little sulphur, natron, and copper. *Rutty. Monro*, vol. I. p. 307.

90. Macroomp, *Cork*. Contains iron, and natron, according to Dr. *Rutty*. A gallon yielded

yielded only eight grains of sediment. *Smith*, vol. II. p. 275. *Monro*, vol. I. p. 308.

91. Mount Pallas, *Cavan*. Contains iron in a volatile acid, with some Epsom salt and calcareous earth. *Rutty*. *Monro*, vol. I. p. 405.

92. Nobler, *Meath*. A gallon is said to yield 170 grains of residuum, most of which, we are told, is vitriol of iron. *Rutty*. *Monro*, vol. I. p. 254.

93. Newton Stewart, *Tyrone*, near Castle hill. Contains iron in a fixt vitriolic acid, some sea-salt, Epsom salt, and earth. *Rutty*. *Monro*, vol. I. p. 370.

94. Rostillan, *Cork*. A pretty strong chalybeate. *Smith*, vol. II. p. 268.

95. Ship Pool, *Cork*, seven miles from the city. A strong chalybeate. *Smith*, vol. II. p. 273.

96. Timoleague, *Cork*. A weak chalybeate. *Smith*, vol. II. p. 271.

97. Tralee, *Kerry*. Contains iron, sea-salt, and absorbent earth. *Rutty*. *Monro*, vol. I. p. 368.

98. Wexford,

98. Wexford. Contains iron, a very little Epsom salt, and some calcareous earth. *Rutty. Monro*, vol. I. p. 281.

Wales.

99. Llandrindod, *Radnor*. Contains a considerable proportion of iron in a volatile acid, and probably a neutral salt. *Linden. Monro*, vol. I. p. 276.

100. Swansea, *Glamorganshire*. A gallon yielded 30 grains of green vitriol, and 8 grains of calcareous earth. *Rutty. Monro*, vol. I. p. 249.

Scotland.

101. Aberbrothock. A weak chalybeate. *Monro*, vol. I. p. 278.

102. Dunse. Contains a little iron, sea-salt, and bittern. *Home. Monro*, vol. I. p. 366.

103. Glendy, *Mairn*. A chalybeate of moderate impregnation. *Monro*, vol. I. p. 278.

104. Hartfell, *Annandale*. A gallon evaporated yields about 30 grains of salt of steel, and 3 or 4 grains of earth. *Med. Essays*, vol. I. art. 12. *Monro*, vol. I. p. 250.

105. Peter-

105. Peterhead, *Aberdeen.* A strong chalybeate. *Monro,* vol. I. p. 278.

ALUMINOUS WATER. Changes vegetable blues red, even after standing some time in the open air; effervesces with alkalies, and is decomposed; precipitating in floculi.

1. Somersham, *Huntingdonshire.* Said to contain alum, iron, calcareous earth, and selenites. *Layard. Phil. Transf.* vol. LVI. *Morris. Monro,* vol. I. p. 432.

WARM WATERS.

1. Bath, *Somersetshire.* Lighter than rain-water. Contains a very small proportion of iron in a volatile vitriolic acid, very little sea-salt, *hepar sulpuris e calce viva*, selenites and fixt air. Raises Fahrenheit's thermometer to 119 degrees. *Lucas. Falconer. Linden. Sutherland. Rutty. Charlton. Hillary. Monro,* vol. II. p. 233.

2. Matlock, *Derbyshire.* Raises Fahrenheit's thermometer to about 70. Lighter than rain-water. Contains a very small proportion of Epsom salt, and calcareous earth. *Short. Monro,* vol. II. p. 261.

3. Bristol,

3. Bristol, *Somersetshire*. Raises Fahrenheit's thermometer to 80. Lighter than rain-water. Contains, in small proportion, Glauber's salt, Epsom salt, calcareous earth, and fixt air. *Lucas. Sutherland. Rutty. Monro*, vol. II. p. 369.

4. Buxton, *Derbyshire*. Raises Fahrenheit's thermometer to 90. Lighter than rain-water. Contains, in small proportion, sea-salt, Epsom salt, and calcareous earth. *Short. Monro*, vol. II. p. 378.

5. Mallow, *Cork*. Raises Fahrenheit's thermometer to 68. Lathers well with soap. Contains a small proportion of sea-salt, Epsom salt, and calcareous earth. *Rutty. Monro*, vol. II. p. 386. *Smith*, vol. II. p. 276.

PETRIFYING WATER. Contains calcareous earth either dissolved in the vitriolic acid in form of selenites, or suspended, in the state of quick-lime, being deprived of its fixt air. If the first, it precipitates with mild or caustic alkalies; if the second, with mild alkali only.

1. Ball or Band Well, *Lincolnshire*, near Henfield. Contains also Epsom salt, and sea-salt. *Short*, vol. I. p. 107. *Monro*, vol. I. p. 464.

2. Cave,

2. Cave, *Fife*, Scotland. *Sibbald, Phil. Tranf. Abridg.* vol. II. p. 325. *Monro*, vol. I. p. 468.

3. Chinkwell, *Dublin*. *Rutty. Monro,* vol. I. p. 470.

4. Glevely, *Rofs*, Scotland. *Monro*, vol. I. p. 468.

5. Hermitage, *Dublin*. Contains alfo a little Epfom falt. *Monro*, vol. I. p. 470.

6. Howth, *Dublin*. Contains alfo fome fea-falt. *Rutty. Monro*, vol. I. p. 469.

7. Llangybi, *Carnarvonfhire*, Wales. Examined by Dr. *Linden* only, from whofe report it is impoffible to know what it contains. *Monro*, vol. I. p. 466.

8. Loggfhiny, *Dublin*. *Rutty. Monro*, vol. I. p. 469.

9. Knarefborough, *Yorkfhire*. Contains, befides a confiderable proportion of earth, a good deal of Epfom fait. *Short*, vol. I. p. 106. *Monro*, vol. I. p. 464.

THE END.

INDEX

INDEX I.

GENERA.

Amphybiolithus	54	Marga	8
Antimonium	40		
Argentum	31	Phytolithus	52
Argilla	1	Plumbum	33
Arsenicum	41		
Aurum	31	Quartzum	18
Bitumen	27	Sal acidum	21
		—- alkali	22
Calx	3	—- metallicum	24
Cobaltum	42	—- neutrum	23
Cuprum	34	—- terreum	25
Entomolithus	56	Saxum	11
		Shistus	14
Ferrum	36	Stannum	32
Fluor	17	Sulphur	29
Graptolithus	57	Talcum	9
Gypsum	8		
		Vismutum	38
Helmintholithus	44		
Humus	1	Zincum	39
		Zoolithus	53
Ichthyolithus	54		

INDEX II.

SPECIES.

Acid of sea-salt	22	Bloodstone	36
— vitriolic	21	Bluish ore	37
Alabaster	9	Bones	53
Alkali volatile	23	Breccia	13
Alum, rock	25	Buccinum	48
Amber	27	Bufonites	55
Anomites	44	Bulla	49
Antimony crystalized	41		
Antimonial ore, red	41		
—————— striated	40	Calamine	39
Arca	51	Chalk	4
Arsenic	25	Chama	50
—— native	41	Clay, common	2
—— white	ib.	—— pipe	ib.
Asbest	10	—— porcelain	ib.
Asteria	46	—— potters	ib.
Astroites	47	Coal	28
Atrion	45	Cockle	50
		Cock-spurs	54
		Cobalt, black	42
Belemnites	46	—— ochre	ib.
Bismuth, native	38	—— ore	ib.
—— flowers of	39	—— crystalline	43
—— ore	ib.	Conglutinated stones	13
Black lead	30	Conus	49
Blend	40	Copper, mundic	35
Bole	2	—— native	34

Copper

Copper nickel	43	Helix	49
—— ore, grey	35	Horsetail	52
—— pyrites, white	ib.		
Cornu ammonis	44	Jasper	19
Crab	56	Jet	28
Crystal	18	Iron ocre	36
Crystalline ore	36	—— ore	37
Cypræa	52	Judaicus	45
Dentalium	48	Lead crystals	33
Dendrites	59	—— glance	33
		—— ocre	34
Earth common	1	—— spar	34
Echinites	45	—— trail	33
Eisenman	37	Leaves of trees	53
Elephants tusks, &c.	53	Lime stone	7
Entrochus	46	Limpet	51
		Load-stone	37
Ferns	52		
Fishes	ib.	Madreporus	47
Flint	18	Magnet	37
Fluor crystallized	17	Manganese	38
Fossile pods	55	Marble	4
Free stone	11	Marcasite	29
Fruit	53	Marle	8
Fullers-earth	3	Millopora	47
		Molares of the sea-wolf	55
		Mountain blue	34
Garnet	19	—— green	35
Glass copper ore	35	Mundic	9, 35
Glossopetræ	54	—— white	42
Glimmer	10, 37	Murex	48
Gold dust	31	Muscle	51
Granite	12	Mya	ib.
Grass	52		
Gryphites	45	Natron	22

Vol. III.   H

Natron		22	Serpent stone	44
			Serpula	48
			Sharks teeth	54
			Shirl	20
Oil fossile		27	Siliquastra	55
—— rock		ib.	Silver mineralized	32
Oister		49	Slate ash-colour	16
Orpiment		41	—— black	14
			—— brown	15
			—— blue	16
Plectronites		54	—— green	15
Pholas		52	—— Irish	17
Plaister earth		9	—— red	15
——— stone		ib.	—— purple	15
Plaister stone fibrous		ib.	Snail	49
			Soap stone	3
			Solen	51
Reeds		52	Spar	7
Roots		53	Stags horns	53
Rotten-stone		3	Star stone	47
			Strombus	49
			Stone marle	8
Sal ammoniac		23		
Sal ammoniac fixt		25	Tellina	50
—— bitter purging		ib.	Tin crystals	32
—— common		23	—— native	ib.
—— Epsom		25	—— stone	ib.
—— Glauber's		23	Toothshell	48
—— Magn. Glaub.		25	Topshell	ib.
Sand		14	Tripoli	3
—— stone		11	Trochus	48
Saxum		11	Tubiporus	47
Screw shell		48	Turbo	48
Scallop		49	Turf	29
Sea star		45	Turcosa	53
Selenites		9	Venus	50

Venus	50	Wood petrefied	52
Vertebra	55		
Vitriol blue	24	Zinc crystallized	39
——— green	ib.	——— ore	40
——— white	ib.		
Voluta	49		

INDEX

INDEX III.

MINERAL WATERS.

A		
Aberbrothock	90	
Acton	62	
Alford	62	
Alkaline water	58	
Alkerton	63	
Aluminous water	91	
Anaduff	71	
Aphaloo	71	
Ardarick	85	
Asheron	66	
Ashwood	71	
Astorp	75	
Aswerby	75	
Athlone	85	

Bagnigge	63, 76
Ball well	63, 92
Ballycastle	85
Bally-mortogh	74
Ballynahinch	71
Ballyspellan	86
Ballyvourney	85
Bandon	85
Barnet	63
Barrowdale	60
Bartholomew's well	59

Bath	90
Bear-forest	85
Bilton	66
Binley	76
Birmingham	75
Bournley	76
Bristol	92
Broughton	66
Buglawton	66
Burlington	75
Buxton	75, 92

Cannock	76
Cape-clear	59
Cargyrle	61
Carlton	76
Carrickfergus	61, 65
Carrickmoor	59
Carrignacurra	86
Carstarphin	73
Cashmore	86
Castleconnel	86
Castlemaign	72
Castle Townshend	86
Cathartic water	62
Cave	93

Cawley

[101]

Cawley	67
Cawthorp	77
Chadlington	67
Chalybeate water	74
Cheltenham	77
Chinkwell	93
Chippenham	77
Clifton	58
Cobham	76
Codfal Wood	67
Coolauran	86
Colurain	76
Comner	63
Copper water	74
Coventry	77
Crickle	67
Croft	67
Cronacree	86
Cronebaun	74
Crofstown	86
Cunley-house	67
Deddington	68
Derby	77
Derrindaff	72
Derryhence	72
Derrylister	72
Dew	57
Dog and Duck	63
Dortshill	60, 77
Drigwell	68
Drumasnave	72
Drumgoon	72
Drumrastle	87
Dulwich	63
Dunnard	87
Dunse	90
Durham	67

Epsom	64
Felstead	77
Filah	78
Five-mile Bridge	87
Gainsborough	68
Galway	87
Garret's Town	87
Glanagarin	87
Glanagarin	87
Glendy	90
Glevely	93
Granshaw	87
Glassonbury	58
Hail	57
Hampstead	78
Hanleys	64, 78
Hartfell	90
Hartlepool	78
Hermitage	93
Harrigate	68, 78
Holt	64
Howth	93
Jessop	78
Ice	57
Ilmington	78
Islington	78
Kanturk	88
Keedleston	68
Kerby	78
Kilindonnel	88
Kilinshanvally	88
Killasher	72

Killasher	—	72	Newton Stewart	89
Killbrew	—	88	Nobler — —	89
Kilpadder	—	83	Normanby —	69
Kilroot	—	62	Northall — —	64
Kinalton	— —	64	Nottington —	69
King's cliff	— —	78		
Knaresborough	—	93	Orston — —	81
Knowsley	— —	78	Owen Breun —	73
Lancaster	— —	80	Pancras — —	65
Landrindod	65, 73,	90	Pelligoe —	73
Latham	— —	80	Petrefying waters	92
Leamington	— —	61		
Leez	— —	80		
Lincomb	—	80	Quin Camel —	69
Lisbeak	—	73		
Lis-done-varna	—	88	Rain water —	57
Llangybi	—	93	Rippon —	69
Loansbury	—	68	River water —	58
Loggshiny	—	93	Road —	81
			Rostillan —	89
Macroomp	—	88	Rougham —	61
Mahereberg	—	62		
Mallow	— —	92	St. Erasmus —	60
Malton	—	81	Salt springs —	61
Malvern	—	80	— water —	60
Markshall	— —	80	Scarborough —	82
Matlock	— —	91	Sea water	60
Maudsley	— —	69	Sene —	82
Mechan	— —	73	Shadwell —	81
Moffat	—	74	Shapmoor —	70
Morton fee	—	64	Shettlewood —	70
Moss-house	—	80	Shippool —	89
Mount Pallas	—	89	Skipton —	70
			Snow —	57
Nevil Holt	— —	65	Somersham —	91
Newham regis	—	81	Spring water —	57

Stanger

[103]

Stanger	——	82
Stenfield	——	82
Stockport	——	82
Stretham	——	65
Sulphur water	—	66
Sutton bog	—	69
Swadlingbar	——	73
Swanſea	——	90
Sydenham	——	65
Tarleton	——	82
Thetford	——	82
Thorparch	——	70
Thornton	——	82
Thurſk	——	82
Tibſhelf	——	82
Tilbury	——	59
Timoleague	——	89
Tober-bony	——	59
Townley	——	82
Tralee	——	89
Tunbridge	——	82
Upminſter	——	70
Wardrew	——	71
Warm waters	——	91
Water pure	——	57
Weatherſlack	——	85
Wellenborough	——	84
Weſt Aſhton	——	84
Weſtwood	——	84
Wexford	——	90
White-acre	——	84
Wigan	——	84
Wigglesworth	——	70
Wirkſworth	——	71
Witham	——	84

www.ingramcontent.com/pod-product-compliance
Lightning Source LLC
Chambersburg PA
CBHW030905170426
43193CB00009BA/744